I0016479

Unlock Your Creativity with Photopea

Edit and retouch images, and create striking text and designs with the free online software

Michael Burton

Unlock Your Creativity with Photopea

Group Product Manager: Rohit Rajkumar

Publishing Product Manager: Chayan Majumdar

Book Project Manager: Aishwarya Mohan

Senior Editor: Nathanya Dias

Technical Editor: Simran Ali

Copy Editor: Safis Editing

Indexer: Hemangini Bari

Production Designer: Jyoti Kadam

DevRel Marketing Coordinators: Anamika Singh and Nivedita Pandey

First published: July 2024

Production reference: 1060624

Published by Packt Publishing Ltd.

Grosvenor House

11 St Paul's Square

Birmingham

B3 1RB, UK.

ISBN 978-1-80181-664-9

www.packtpub.com

I would like to dedicate this book to Michael and Madelyn, and my wife Tamela. You all played a major role in helping me get this book finished.

I would also like to thank my wife for her patience and understanding while I put in long hours to complete this book. There were definitely a lot of family responsibilities. Lastly, I would also like to thank my parents, brothers, sister, and my willing cousin, for assisting me with images and all their genuine support.

–Michael Burton

Foreword

From the moment I first encountered Michael, just before his graduation from Southern New Hampshire University with a BA in graphic design and media arts in 2021, I was utterly captivated by his remarkable blend of creativity and versatility. As a senior marketing consultant and world illustration judge, I have witnessed countless examples of artistic prowess, but Michael's work truly stands out. His seamless integration of illustration, graphic design, branding, and music into multimedia projects is nothing short of exhilarating—it harmonizes storytelling, brand elevation, and compelling selling points in a way that captivates and engages.

Upon recognizing the award-winning potential of his creations, I urged him to share his genius with the world. The announcement of his forthcoming book, *Unlock Your Creativity with Photopea*, confirmed my belief that this was the ideal platform for Michael to showcase his vast knowledge and inventive flair as a multimedia artist.

Observing Michael's strategic repositioning since our impactful discussion in 2021 has been a profound testament to his unyielding passion for learning, sharing, and inspiring. His work now resonates with renowned music artists, public figures, and influencers across the globe, creating a rich tapestry of collaboration and creativity.

Readers of his book are in for a treat. They will journey through a meticulous step-by-step guide on mastering Photopea—from navigating its intricate workspace and understanding the photo retouching fundamentals to exploring logo design, drawing, and painting. Michael also offers a deep dive into his methodical approach to idea development, and so much more.

I wish Michael the very best as he continues to inspire and innovate. I know that every reader will be as enchanted and motivated by his creativity as I have been.

Ellie Altomare

Senior Marketing Consultant and World Illustration Judge

Contributors

About the author

Michael Burton is an experienced graphic artist, who evolved into multimedia. Since 2000, he's worked in branding and decorated apparel, and he uses Photoshop and alternative software for digital imaging, screen-printing, drawing, and painting. He has designed for hundreds of schools and local businesses, including Chicago Public Schools and the Illinois High School Association. Some of his college and pro-league clients include WNBA player Candace Parker and Brian Urlacher of the Chicago Bears. In addition, he works in vector art, loves creative writing, music, video editing, and spoken word for graphic storytelling, and worked with a 4x Grammy-winning artist and author of the creative memoir, *Let Me Paint a Picture*. He has earned two associates and a BA in graphic design and media arts at Southern New Hampshire University.

I would also like to thank Urvi Shah for presenting the opportunity to write this book.

About the reviewer

Jamila Surpris is an environmental scientist and graphic designer residing in the Mvskoke (Atlanta, Georgia). She is passionate about problem solving by bridging gaps between STEM and creative fields. Her work primarily supports nonprofit and mission-based organizations through brand development and web design. She has collaborated on other books such as *Unbelonging* by Gayatri Sethi, as a first reader and collateral branding designer. Social responsibility, equity, decolonial principles, and human-centered design are crucial points of her work. Jamila is especially excited to continue uplifting other Black, Brown, Indigenous, and Queer organizations and individuals through her work and collaborations as an independent designer.

Table of Contents

Part 2: Digital Imaging, Design Techniques, and Other Software

6

7

8

Part 3: Drawing Figures, Creating a Logo, and Other Features

12

Advanced Color Techniques   321

13

Bonus: How to Draw and Paint a Figure and Character   357

14

Bonus: How to Create a Logo   387

15

Tips, Tricks, and Best Practices   411

Preface

Welcome to *Unlock Your Creativity with Photopea*!

You'll learn the essentials for photo editing and design, layer effects, retouching, and arranging images with enticing text, as well as the overall fundamentals that pertain to any photo editing software in a layered step-by-step process. Each essential task will build up your skillset to execute features as they gradually advance. This will create an organized, fun, and effective way of learning how to use Photopea.

Photopea has gained popularity over the last several years, as a free open-source image editing application, packed with amazing features. As an experienced graphic designer evolving to multimedia, I have worked in a number of different software applications over the years, and was honored to have the opportunity to share some of my education and work experience related to image editing, illustration, and so on, in Photopea.

I had many obstacles to overcome in trying to get this publication complete. These included moving to a new house the same week I was scheduled to get started on the book, adjusting to a work schedule that is the opposite of the 9-5, being a father of twin toddlers, and other things I couldn't make up... the list goes on.

Anyway, I'm glad I pressed through to complete the book, and hope that you find it helpful and inspirational!

Who this book is for

This book is for photographers, illustrators, graphic designers, hobbyists, and students in need of alternative software suitable for digital imaging and design. Newbies can grasp the fundamentals of photo editing with no prior knowledge but would benefit from following the step-by-step tutorials.

What this book covers

Chapter 1, *Taking your Design and Editing to the Next Level with Photopea*, shows the advantages, power, and flexibility of being able to use Photopea from any device or browser for free. We'll also cover the fundamentals of working in different modes and file types.

Chapter 2, *Creating and Working with Image Editing Documents in Photopea*, looks at different document types, sizes, templates, guides, snapping, using artboards, settings for output printer quality for web and physical print, arranging and aligning elements within the document, and so on.

Chapter 3, Navigating and Using the Workspace, involves navigating the workspace, explaining the purpose and features of the tools in the toolbar, and using the top menu bar and flyout menus to access and execute tasks for creating and saving documents, applying effects, editing type, and so on.

Chapter 4, Layer Management: Creating, Organizing, and Applying Advanced Features, covers layers and the various uses and features you can apply to each layer. Starting with creating, naming, and organizing layers to grouping layers into folders, understanding and creating masks, applying layer styles, and more.

Chapter 5, Understanding Selection Fundamentals, looks at different selection styles and the selection tools involved in the process, including lasso, polygon lasso, rectangle, elliptical selects, and the pen tool. We will also cover the Refine Edge tool and working with channels.

Chapter 6, Color Theory and Application, covers exploring and understanding the basics of color theory, design concepts, and principles that are involved in creating great compositions. We will also look at using and creating swatches, choosing color profiles, using the color picker, and more in Photopea.

Chapter 7, Using and Creating Brushes, starts with a solid overview of what the brushes are capable of achieving in your projects. In addition to that, we will look at the Brush Panels, Presets, and Patterns, as well as how to import and create our own brushes.

Chapter 8, Photo Retouching Techniques, breaks down, step by step, how to touch up a headshot, adjust hair, and restore an old photo. This will be achieved with the basic tools: the brush tool, clone stamp, eraser, pencil tools, layers, and masks.

Chapter 9, Exploring Advanced Image Compositing Techniques, combines all the things we've been learning: making adjustments to images with lighting, retouching, selecting, combining, and arranging various objects, elements and backgrounds into a single believable image composite.

Chapter 10, Text Fundamentals and Styling in Photopea, looks at the foundation of text, choosing the right font and typography, using the text tool for Artistic Text, warped, and shaped text; along with body text for paragraphs and large bodies of text that will often use character styles.

Chapter 11, Pre-Designed Templates, Extra Effects, and Features Overview, explores the resourceful library of pre-designed templates in Photopea. We will customize some for social media posts, create an animated logo, and look at the Vectorize a Bitmap feature.

Chapter 12, Advanced Color Techniques, will dive into a sub-area on advanced color techniques. It will give you a better understanding of what channels are, how to work in channels, and how to add color to black-and-white images.

Chapter 13, Bonus: How to Draw and Paint a Figure and Character, will explore ways of getting comfortable with drawing, creating rough sketches, and polishing them up to a final drawing. Afterwards, we prepare the final drawing as a grayscale value study, and finish the painting in full color.

Chapter 14, *Bonus: How to Create a Logo*, begins with understanding what a logo is and how it is used. Afterwards, we cover how to create a logo. This will include brainstorming, sketching out ideas, finishing a logo in black and white, and finally, adding color to finish it.

Chapter 15, *Tips, Tricks, and Best Practices*, involves understanding smart objects and other features of Photopea that you may not be aware of, such as working with the Vanishing Point Filter features, Layer Comps, and Building a Portfolio for a career.

To get the most out of this book

Be consistent with following along with the step-by-step processes for different design projects. Invest in a drawing tablet to execute some of the more challenging projects, and make sure you have a good internet service provider to make sure Photopea runs smoothly.

Software/hardware covered in the book	Operating system requirements
Photopea	Windows, macOS, or any smart device
Affinity Designer	Windows or macOS
Character Creator 3	Windows or macOS

> **Important note**
> The book will be accompanied by a resources folder titled `Unlock Your Creativity Resources`. It will include the original photos and Photopea documents. *Chapters 4-8* include the photos. *Chapters 9-15* include the photos and Photopea files.

Download the exercise files

You can download the exercise files for this book at `https://packt.link/gbz/9781801816649`

Conventions used

There are a number of text conventions used throughout this book.

`Code in text`: Indicates code words in text, database table names, folder names, filenames, file extensions, pathnames, dummy URLs, user input, and Twitter handles. Here is an example: "Now that the *MIC* logo is finished, we can export the vectorized bitmap logo as an `SVG` or `PDF`, or as a `Vector` file that can be opened in other vector-based applications and software such as Adobe Illustrator, Corel Draw, Affinity Photo, and Inkscape, among others."

Bold: Indicates a new term, an important word, or words that you see onscreen. For instance, words in menus or dialog boxes appear in **bold**. Here is an example: "Next, while the image is selected (with the **Selection** Tool), go to the **Edit** menu, choose **Define New**, and select **Brush**."

> **Tips or important notes**
> Appear like this.

Get in touch

Feedback from our readers is always welcome.

General feedback: If you have questions about any aspect of this book, email us at customercare@packtpub.com and mention the book title in the subject of your message.

Errata: Although we have taken every care to ensure the accuracy of our content, mistakes do happen. If you have found a mistake in this book, we would be grateful if you would report this to us. Please visit www.packtpub.com/support/errata and fill in the form.

Piracy: If you come across any illegal copies of our works in any form on the internet, we would be grateful if you would provide us with the location address or website name. Please contact us at copyright@packtpub.com with a link to the material.

If you are interested in becoming an author: If there is a topic that you have expertise in and you are interested in either writing or contributing to a book, please visit authors.packtpub.com.

Share Your Thoughts

Once you've read *Unlock Your Creativity with Photopea*, we'd love to hear your thoughts! Scan the QR code below to go straight to the Amazon review page for this book and share your feedback.

https://packt.link/r/1-801-81664-6

Your review is important to us and the tech community and will help us make sure we're delivering excellent quality content.

Download a free PDF copy of this book

Thanks for purchasing this book!

Do you like to read on the go but are unable to carry your print books everywhere?

Is your eBook purchase not compatible with the device of your choice?

Don't worry, now with every Packt book you get a DRM-free PDF version of that book at no cost.

Read anywhere, any place, on any device. Search, copy, and paste code from your favorite technical books directly into your application.

The perks don't stop there, you can get exclusive access to discounts, newsletters, and great free content in your inbox daily

Follow these simple steps to get the benefits:

1. Scan the QR code or visit the link below

https://packt.link/free-ebook/9781801816649

2. Submit your proof of purchase

3. That's it! We'll send your free PDF and other benefits to your email directly

Part 1: Getting Started With Photopea

In this part, we will cover creating and working with documents in Photopea. We will get comfortable with navigating and utilizing the workspace. We will also learn the importance of layer management, that is, creating, organizing, and applying advanced features. Lastly, we will gain an understanding of the fundamentals of selection.

This part comprises the following chapters:

- *Chapter 1, Taking Your Design and Editing to the Next Level with Photopea*
- *Chapter 2, Creating and Working with Image Editing Documents in Photopea*
- *Chapter 3, Navigating and Utilizing the Workspace*
- *Chapter 4, Layer Management: Creating, Organizing, and Applying Advanced Features*
- *Chapter 5, Understanding Selection Fundamentals*

1

Taking Your Design and Editing to the Next Level with Photopea

This book will teach you the essentials of photo editing and design, which includes: layer effects, retouching, arranging images with enticing text in a step by step process. Each essential task will build up your skillset to execute features as they gradually advance. This will create an organized, fun, and effective way of learning Photopea, and can be applied to most photo editing software.

There are several advantages of using Photopea:

- Firstly, it doesn't take as much computing power as other photo editing programs.
- It's free to try, and free to use as often as you'd like without watermarks on your documents.
- Secondly, it can run on all your devices (including smartphones), but I would advise working on a large screen tablet, laptop, or desktop when working to easily navigate in the interface.

The only issue that can occur is that it may run slow at times depending on how strong the internet connection is.

In this chapter, we will cover how to easily create an account to use Photopea and work directly in the web browser, how to download and access it from your desktop and compare the free versus paid version.

From there, we will learn the about raster and vector images, when to use raster and vector, and break down the different Color Modes; RGB versus CMYK, and greyscale.

In this chapter, we will learn the following skills:

- Learn how to set up a user account and properly install the software.
- Learn about the difference between the free version and paid subscription; help users decide which option is best for them.
- Learn what raster and vector images are, and how and when to use them.
- Learn about RGB and CMYK when using color for documents, monitors, and setting up documents for different printers.

In this chapter we will cover the following topics:

- Installing Photopea and Setting up an Account
- Free versus Premium Version
- Raster and Vector Images Explained
- RGB versus CMYK Color Modes

Technical requirements

We will be using the Photopea Photo editing software for all of the projects in this book. You can work with Photopea directly in your web browser by clicking on the link below: (`https://www.photopea.com`).

You also have the option to download and install Photopea to your Desktop. (recommended)

Installing Photopea and Setting up an Account

Now that we understand the benefits and technical requirements using Photopea, we can setup up an account, in addition, we can install it on the desktop. (Neither option is required, you can start working on a new document immediately).

1. To setup an account for Photopea, go to (`https://www.photopea.com`) select Account, choose free version, and login with your email, (see *Figure 1.1*).
2. Next, select which platform you prefer to sign-in with. I chose my Gmail email account to sign-up, and started working directly in my web browser.

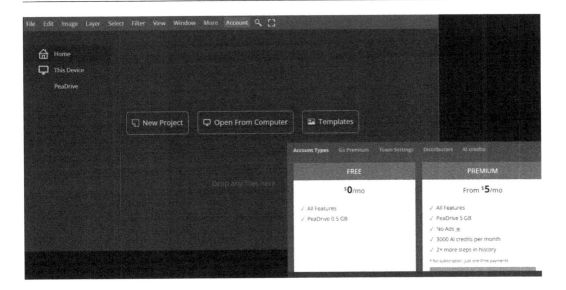

Figure 1.1 – Setting up an Account

So, in the previous steps, we created an account on Photopea, and understand that we can work and access it from the browser, as well as open and work on our projects from any computer with our registered account.

Next, we will learn how to install it to work on a desktop computer. This will enable to use Photopea with more power, faster response, and less delays, since we're not totally dependent on a browser to operate it; but we still need to be connected to the internet to run the software.

To access Photopea from your desktop, do the following:

1. Open Google Chrome and type (`https://www.photopea.com/`).

2. The Photopea app will load in the web browser.

3. Click on *More* at the top of the Photopea Menu bar and select **Install Photopea**.

4. Once Photopea is installed, it will create a shortcut icon on your desktop screen.

5. *Right-click* on the Photopea shortcut icon and choose, **Pin to Start**, or **Pin to Taskbar**. This makes it is easy to access Photopea and begin working..

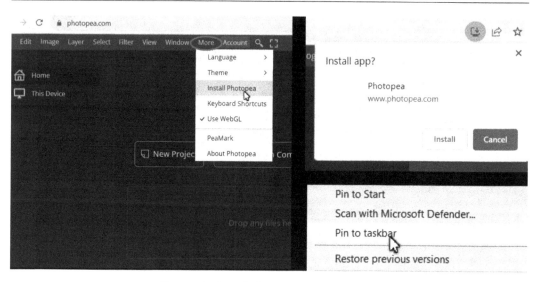

Figure 1.2 – Install Photopea on the Desktop

6. If you need to Uninstall Photopea, type in the google chrome apps link in your browser: (chrome://apps/).

7. You will see all of the Chrome apps installed. *Right-click* over Photopea and *select* **Uninstall**:

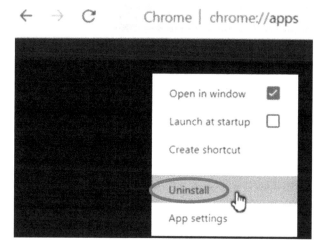

Figure 1.3 – Uninstall Photopea

Now that we've installed Photopea to the desktop, and learned how to **Uninstall** It; we can move on to the next section.

Free versus Premium Version

The free version of Photopea has pop-up ads, but a recent update allows you to hide them in Google Chrome Extensions called, **Remove Ads from Photopea**. The free version includes most features and 0.5 GB of cloud storage. That's great news for new users to get started risk-free and good for experienced users on a budget.

To remove ads from Photopea, do the following:

1. Open **GOOGLE Chrome** and type in `https://chromewebstore.google.com/`.

2. Type Photopea in the *Search extensions and themes bar*, and select the **Remove Ads from Photopea** extension.

3. Next, click on the **Add to Chrome** button, located on the right side. (see *Figure 1.4*):

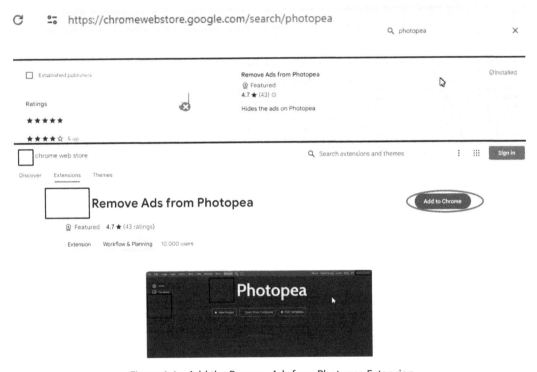

Figure 1.4 – Add the Remove Ads from Photopea Extension

Important note

The remove ads feature only works in the Free Version. If you pay for the Premium version, you have to go back in the Chrome Extensions to Disable it. Otherwise, the Premium Version will act buggy and not display right. This is because the Premium version has a built in extension that removes the ads already.

The premium version is $5/month with 5GB cloud storage, comes with no ads, quick email support, saves more of the steps you made working in the document (in the history tab), and now includes the new AI Selection Feature that we will cover in *Chapter 15, Tips, Tricks, and Best Practices*.

> **Important note**
>
> The **History** Tab allows you to revisit and recover an original action you made.

I will be sticking with the free version of Photopea for a couple of reasons.

1. I can hide the Ads in the Google Chrome Extensions. Go to the top right of the Google Chrome Browser, click on the 3 small dots, scroll halfway down, select **Extensions**, **Photopea, Remove the Ads** and **Hide the ads** on Photopea). (see *Figure 1.5*).

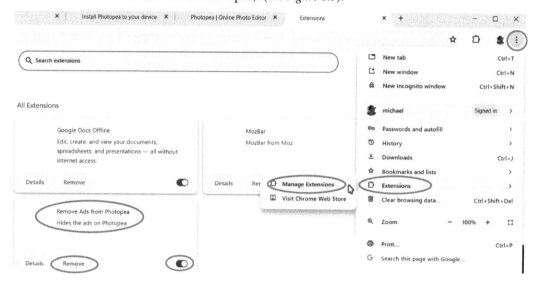

Figure 1.5 – Remove or Disable the Remove Ads from Photopea Extension

2. I will use my personal external hard drives along with the free Google cloud drive to back up my files.

Next, we will take a look at Raster and Vector Images, how they differ, and will learn to choose the best format for working on projects and jobs.

Raster and Vector Images Explained

Raster and Vector are the two most common formats used for image editing, logo design, graphic art, animation, printing, and so on. Each format has key distinctions that are important to know before starting a project.

Looking into Raster

Raster files are created from tiny individual rectangular color dots called Pixels. The more pixels you have, the better the quality and details you can add to an image or digital painting. Specific file types such as PSD, JPEG, PNG, BMP, and so on, can handle a lesser or greater number of pixels. View the pixel examples below, see *Figure 1.6*:

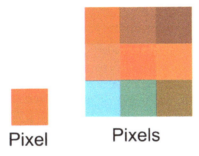

Figure 1.6 – Pixels

Raster images can achieve complex renderings that look natural, soft-blended, multi-colored gradations; and also, mid-tones, lights, and shadows. They are also great for photography and digital painting, and for blending colors. Programs like Photopea, Photoshop, Gimp, and Affinity Photo are ideal for creating and editing raster work.

One of the limitations of raster files is that it is difficult to enlarge images; or scale them down as that might cause them to lose quality or become pixelated. The larger you set up your document and resolution, the larger the file size becomes, and will require more computing power to work, that may cause your computer to run slower. I will explain the importance of DPI or Dots Per Inch in relation to file size, pixels, printing and viewing the quality of documents later on

Looking into Vector

Vector images are made up of paths created by individual nodes using mathematical formulas. Nodes are seen in lines, curves, or points (as wireframes) to make individual vector shapes with or without color. Below is an example of nodes and wireframes. See *Figure 1.7*:

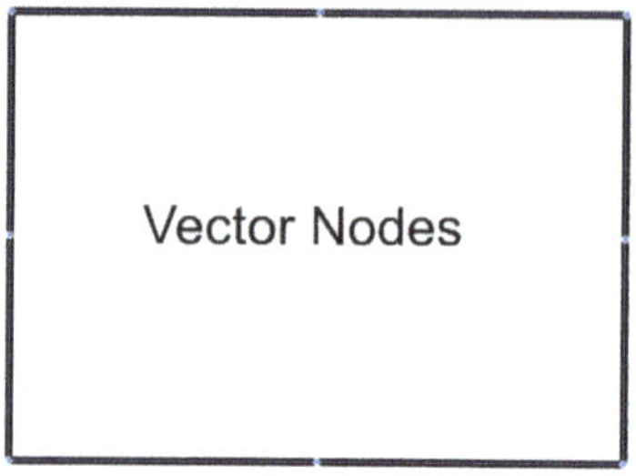

Vector Nodes

Figure 1.7 – Vector Nodes

Vector images and graphics can be infinitely scaled up or down, they are usually smaller file sizes than hi-res raster files; and are best for spot colors, logos, technical drawings, 3D, 2D animation, specialty printing, embroidery, color separations, and CAD drawing. Some popular vector file formats are AI, EPS, PDF, CDR, SVG, AD or the file initials of a vector program not listed.

Vector images are capable of mimicking some rendering detailed work with highlights, shadows, and blending with limitations compared to raster rendering, and would require a lot more effort and time to simulate a close interpretation of a raster image. In order for Vector images to display correctly on the web, they need to be exported into a raster format. Popular programs like Adobe Illustrator, Corel Draw, Affinity Designer, and Inkscape are great for vector-based projects.

Should you use Raster or Vector for your project?

Keep the things we went over about Raster and Vector formats in mind to help make the best decision on choosing when to work in a vector or raster image. If you need infinite scaling use vector, if you need complex, detailed subtle gradations and shadows, choose raster. You can also export vector images into raster programs and add complex raster rendering, and you can open raster files in vector programs to add vector elements.

You can see Raster versus Vector examples in *Figure 1.8*. Notice how fuzzy and pixelated a low-quality raster image is on the far left:

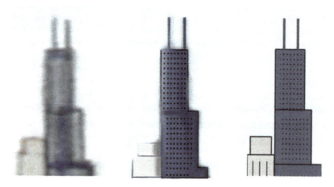

Figure 1.8 – Raster versus Vector (Far left, Raster low resolution) (Vector on far right)

In closing, we learned the difference between Raster and Vector images, when you should work in one format over the other, scaling up, scaling down, and image quality.

We will cover RGB and CMYK color modes in the upcoming section.

RGB versus CMYK Color Modes

Working in Photopea or any Vector or Raster program, you should be familiar with the two common color modes **Red, Green, Blue (RGB)** and **Cayan, Magenta, Yellow, Black (CMYK)**. Photopea has four color profiles but we will only use two color modes, (Photoshop, Affinity Photo and others have even more color modes.) Both have distinctive differences and roles for viewing, mixing, and printing colors. Being up to speed with both modes will reduce surprising unwanted results, costly printouts, and other issues that can occur.

You can change the document from RGB to CMYK by simply clicking on the Image menu, Mode, and choose CMYK or RGB, see *Figure 1.9*:

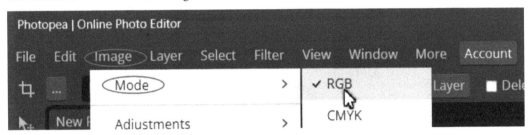

Figure 1.9 – RBG Color and CMYK Color

Let's go over the differences between RGB and CMYK.

RGB

RGB is used for displaying web pages, videos, social media images and graphics elements. *RGB* stands for *Red*, *Green*, and *Blue*, and produces millions of colors by using light that can display up to 16.7 million colors on computers, TVs, smart phones, and any screen device through additive mixing. This is done by displaying red, green and blue bulbs at various intensities similar to the human eye color receptors called cones. The color white is displayed when all three colors are at full saturation, whilst 0 intensity will result in black.

CMYK

CMYK stands for **Cyan**, **Magenta**, **Yellow**, *and* **Black**. CMYK can print around 16,000 colors. All 4 colors can be mixed to create black. In the past, most printing companies required you to export your files as CMYK. The commercial printers would print products like business cards, stationary, booklets, posters, T-shirts, any advertisements for printing, and packaging through subtractive mixing.

There are times when you may have to enlarge the document dimensions to fit on T-Shirts, Banners, Large Posters, etc. some files may become larger than initially created. CMYK also carries more data than vector spot colors.

In today's technology, commercial printing is less reliant on the CMYK process. Nowadays, printers, are equipped with 10+ different color cartridges, and some can even print the color white. Many of my creative friends that work in the field have been recommended not to convert their files to CMYK, let the print operators decide on if a conversion to CMYK is needed. It's best to work and create your documents in RGB; if you need to convert it to CMYK, it's best to Export as CMYK.

Let's look at the example of the female character side by side. The document on the let is an RGB file. The Image on the right is CMYK. If you look closely, the image on the right has more vibrant colors (as it should, being an RGB file). The image on the right has a washed out dull looking blue (as it should being a CMYK format), see *Figure 1.10*:

Figure 1.10 – Comparing an RBG Document vs a CMYK Document

Now let's take a look at the Color Range in the Color Picker example. The **Color Picker** on the left shows all the colors available from the yellow we sampled. Once you click on the CMYK gamut, you see this greyed out shape on the color picker. This is showing colors unavailable in CMYK, see *Figure 1.11*:

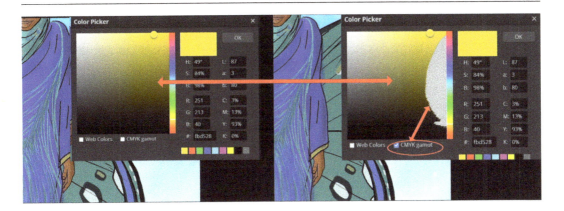

Figure 1.11 – The Colors available in RGB vs CMYK in the Color Picker

That will cover the basics on RGB vs CMYK.

In closing, we have covered the differences between RGB vs CMYK color modes, gained a better understanding of the capabilities of each format, and when you need to setup each mode for printing vs viewing images on your computers, and smart phone devices.

Summary

In this chapter we covered how to setup a user account and properly install software, the difference between the free version and paid subscription, learned the difference between raster and vector images, and also how RGB colors differ from CMYK, and how resolution quality can affect the way images display on monitors, and print out quality from printers.

In the next chapter we will learn how to set-up documents, different document types, sizes, templates, using guidelines, and artboards.

2

Creating and Working with Image Editing Documents in Photopea

In this chapter, we will look at different ways to open a document, how to save a document in Photopea, set up document sizes, use templates with preset sizes, and configure guides, rulers, and snapping to lay out your images and elements in a balanced composition. We will review the settings for output printer quality for web and physical print, and will also get an introduction to artboards for managing multiple pages within a single document – a common feature in programs such as Figma and Sketch and for brainstorming ideas (which we will dive deeper into later in the book).

This chapter covers the following topics:

- Creating and saving new documents and templates
- Opening existing Documents for projects
- Saving documents in Photopea
- Image size and resizing documents for projects
- Guides, rulers, grids, and snapping
- Utilizing artboards

Creating and saving new documents and templates

When creating a new document, it's good to have an idea of what you are creating the document for. This is important because there may be a specific size for a specific type of document required by a printing company, such as an 11"x17" poster, 8.5" x 11" letterhead size, a 5"x7" postcard, and so on. Should it be in vertical or horizontal format? Once you know what you want to do, Photopea makes it easy to get started.

1. For a moment, let's look around the home screen that appears when you launch Photopea. You will notice the **New Project** tab just above the rectangular box, **Drag any files here** on the home screen. To open a new document, we can either click on **New Project** or click on the **File** menu. Let's click on the **New Project** button. See *Figure 2.1*:

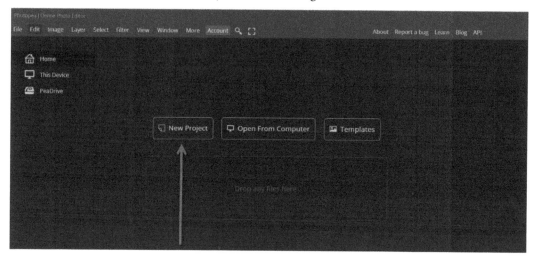

Figure 2.1 – New Project button

2. On clicking **New Project**, we can fill out the following properties for our new project:

    • **Name**, **Width**, **Height**, and **Dots Per Inch** (**DPI**) (which can be changed anytime, as we will do periodically throughout the book)

    • Document background color

    • Options for specific document-sized templates created for social media (for example, a Facebook cover page, YouTube thumbnails, Instagram Stories, and so forth)

    • In addition, we have a choice of templates for printing high-quality files based on common photo sizes, computer screens, and other screen devices such as smartphones and tablets

3. You can follow along with me and select an 11 x 17 poster at 300 DPI for a high-quality printed image. Notice the options for the document, shown in *Figure 2.2*.

4. Change the file name to 11 x 17 poster for now. Switch the **Width** and **Height** measurements to inches, leave the background white, select the **Print** document type, and then select **Ledger (11 x 17 in)** and click **Create**.

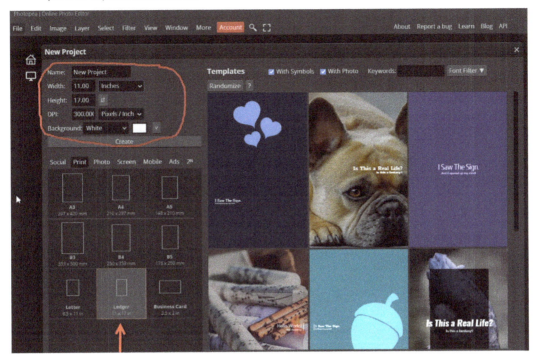

Figure 2.2 – New Project options

5. Now the document is open, go to **Image | Mode**, and you should see **RGB** selected as the file type (see *Figure 2.3*). This is correct for working on the document, but keep in mind that printing companies may require you to convert the file to CMYK and export it as a specific file type (JPEG, PDF, etc.) for their printing process. Just make sure to keep the original file saved as **RGB** for your personal backup. This will keep all the color information accurate for viewing on the web and screen-based devices.

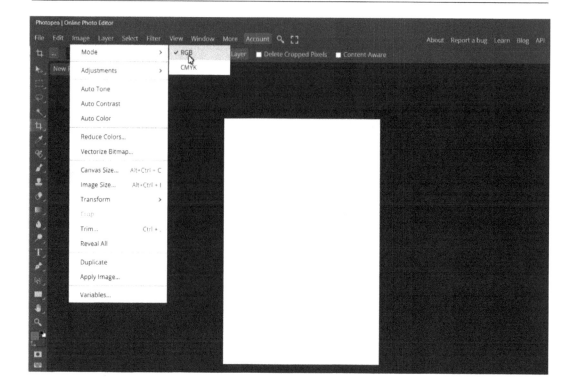

Figure 2.3 – RGB color mode

> **Important note**
>
> Since you are creating a poster for a hi-res printer, you may or may not be able to print to the edge of the paper. Some printers may have a margin leaving a white paper border around the edge.
>
> Keep this in mind for future printing projects and find out ahead of time. Some printing companies may provide a document template set up for you to print to the edges (called a bleed area, from which you want to keep text and any other important design information away). The bleed area is usually best for a background color that you want to cover the rest of the document, and don't mind getting cut off. We will explain more on this later in the book.

That brings us to the end of setting up a new document. We are ready for the next section on opening existing documents.

Opening existing documents for projects

There are several ways to open an existing file in Photopea. Firstly, you can choose **Open From Computer** on the Photopea home screen. Secondly, you can go to the top menu bar and select **File | Open**, or you can use the shortcut keys *CTRL + O* and locate the image or supporting file on your computer (see *Figure 2.4*):

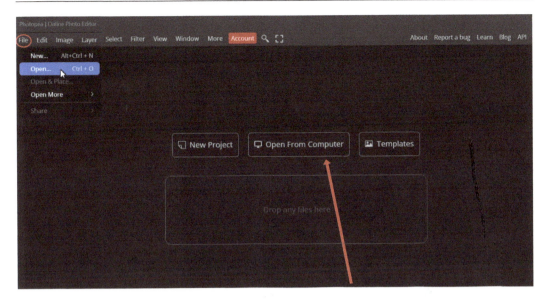

Figure 2.4 – Open a document

Thirdly, you can drag a file or image directly into Photopea's document area and it will open a document in the size of the image (see *Figure 2.5*):

Figure 2.5 – Drag an image directly from a folder

Fourthly, you can drag an existing image into an already-open file and it will appear as another layer within the existing document (see *Figure. 2.6*):

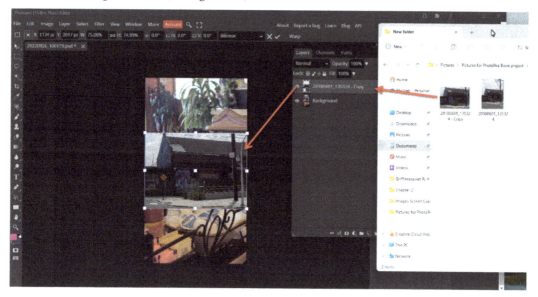

Figure 2.6 – Drag an existing image into an already-open file

Or you can drag a file into the main tab area of an existing file, and it will open in a separate document panel window (see *Figure 2.7*):

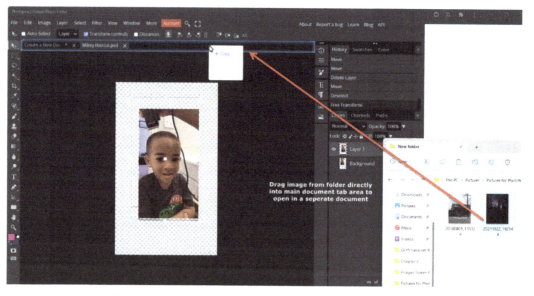

Figure 2.7 – Drag a file into the main tab area of an existing file

Photopea can also open images and files from the **Open More** tab as seen in *Figure 2.8*. This gives you additional options to open an image beyond your local desktop, such as a web browser or Photopea's **Storage** (online cloud), and will ask permission to access your computer. Thirdly, you can take a picture using the **Take a Picture** option from your camera directly from Photopea software.

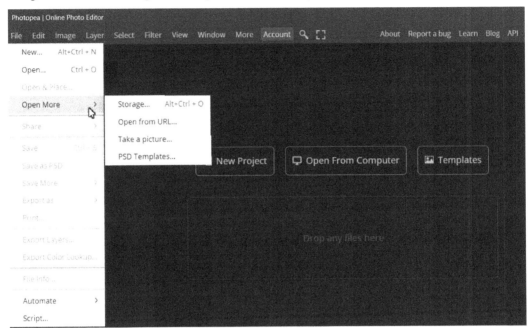

Figure 2.8 – Open More button

Now that we know how to open existing documents, let's move on to saving our work in Photopea.

Saving documents in Photopea

It's important to save your files regularly to avoid losing information and the progress you've made. Photopea gives us different ways to save files in different file types, on the local hard drive, and Photopea cloud. It's important to both understand and utilize the save features so you don't lose your file or any recent edits you've made.

You can access the save features via the top-left **File** menu (see *Figure 2.9*):

- Open the **File** menu and you will see **Save**, **Save (Smart Object)**, **Save as PSD**, and **Save More**. Just below that is an **Export as** option, another save feature.

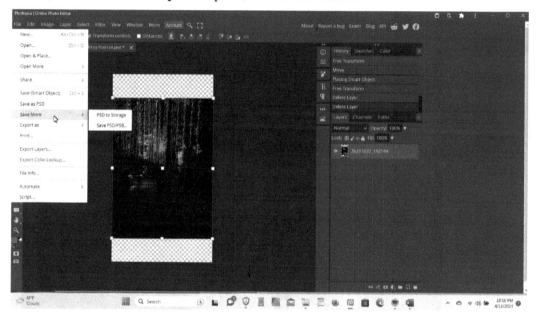

Figure 2.9 – Save features

- Saving a file will only save the document in the cloud (not on your local computer).

- **Save (Smart Object)** is used for a specific function and type of layer.

- **Save as PSD** will save your file to your computer with all the working layers and effects.

- **Save More | PSD to Storage** will save your file with all working layers on Photopea's online cloud (the storage can reach capacity quickly since PSD files are much larger than JPEGs and other flattened files).

- **Export as** allows you to export your document into 16 different file types that we will go over later in the book.

Going to **File** | **Export** | **More** gives you additional file formats to export and save to. See *Figure 2.10*:

Figure 2.10 – Export as file options

This sums up the different ways to save and export files in Photopea. We can go to the next section.

Image size and resizing Documents (canvases) for projects

In Photopea, you can resize your original images. However, you have to be careful while saving! Once you close the **Document** (**Canvas**), you won't be able to revert the file back to its original state. It's best to back up the images in case you need to have the originals.

> **Important note**
>
> There is a clear distinction between changing the **Image Size** versus changing the **Canvas** (or Document) Size. Using the *Transform* and *Scale* option with an image on a separate **Layer** or a single **Layer** *Unlocked*, you can change the **Image Size** without affecting the actual **Document Size** (Canvas). If you try to change the Image Size from the **Image** menu, this option will affect the entire **Document Size (Canvas Size)** as demonstrated in *Figure 2.13*.

Let's learn more about these features in the next section.

Changing an image size

There are a number of ways that you can resize an image:

1. First, *select* the **Layer** of the image you want to scale, with the **Move Tool** in the **Tool Bar** on the top left corner to activate the white dots around the image. Now you can drag the **Move** tool left, or right, up, or down, and it will **Scale** the image up or down proportionally by default.

2. A *second* option while the image is selected, *press* the shortcut keys *Alt + Ctrl + T* to activate the **Free Transform** mode; which allows you to *Rotate* the image, or *Resize* the image proportionally. To do this, gently hover the mouse near the edge of any one of the anchor points, until you see a *diagonal double* pointed arrow.

3. Next, *press, hold and drag the mouse* in or out, or up and down, to increase or decrease the size of the image.

4. A *third* option while the image is active with the **Move** tool, you will see a **Transform controls** option located just below the **Filter** menu. This allows you to *Scale* or *Rotate* the image quickly.

5. A *fourth* way to change the size is to go to the **Edit** menu, scroll down to **Transform**, and *select Scale* to activate the *Transform mode*, and *press, hold,* and *drag* the mouse to increase or decrease the size of the image.

6. A *fifth* option is when you *press* and *hold* the *Shift* Key, or the *Ctrl* Key while dragging the mouse, the image will **Resize** unproportionally. Which will result in a stretched, or squashed image. You may need to use this technique intentionally for a unique design, or something for a surreal image, and so on. See *Figure 2.11*:

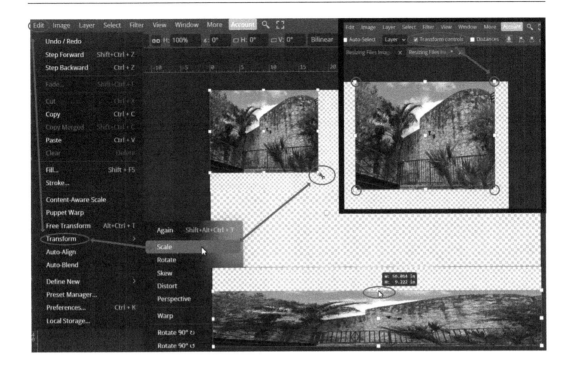

Figure 2.11 – Change an Image Size

7.   A *sixth* option is resizing an image disproportionately, and skewing perspective. To achieve this, press the *Alt + Ctrl + T* keys to activate the **Free Transform** mode.

8.   Hover the mouse near one of the anchor points, then *press* and *hold down Ctrl + Shift* and *left mouse button*.

9.   Drag the mouse in a different direction to skew the image. See *Figure 2.12*:

Figure 2.12 – Resizing an Image Disproportionately, and Skewing Perspective

10. The *seventh* option is simple Go to the top menu bar and select the **Image** menu and scroll
down to **Image Size**. You will see a small square image panel with the **Width**, **Height**, and
DPI options to adjust the size with numerical values to **Scale** up or **Scale** down the **Canvas**
with the images intact. Also, the small chain link icon controls scaling down proportional and
disproportional by clicking the link on and off (see *Figure 2.13*):

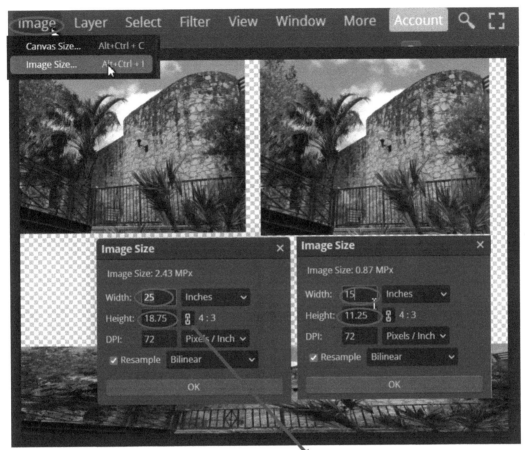

Click to lock in scale to proportion.
Click to unlock to scale disproportion

Figure 2.13 – Adjust image size from the Image menu

That wraps up this section on adjusting the image size. Next, we'll take a look at resizing documents.

Resizing Documents (canvases)

There are a couple ways that you can resize a Document (**Canvas**)

One thing to be aware of is how resizing a document can affect quality of the image and the DPI and pixel size (we will cover those things later on). If made too small, the end result can be low quality, but equally if made too large, it will be impossible to send the document by email or upload to a printing platform, and could even crash your computer.

You can change the Document Size (**Canvas Size**) by navigating to the top menu bar, selecting the **Image** menu, and clicking **Canvas Size**. A small square **Canvas Size** panel with options for scaling and positioning the current document appears. Notice how moving the **Anchor Point** in the **Canvas Size** panel effects the area of the Size results. You can see how the right half of the Canvas got cropped off when I changed the location of the Anchor to the top-left. See *Figure 2.14*:

Figure 2.14 – Canvas Size

Using the **Crop tool** (see *Figure 2.15*) is another way to change the **Canvas Size**. You can crop out a particular area of a photo or image for a better composition, or if you need to leave out some unwanted object or figure you want to remove. We will discuss the **Crop tool** when we go over all the tools in *Chapter 3, Navigating and Using the Workspace*.

Figure 2.15 – Using the Crop tool to change the document size

We have reached the end of *Resizing Documents (Canvases) Image Size and, Crop Tool* section. Let's move on to the next section.

Guides, rulers, grids, and snapping

As you gradually unlock your creativity with each chapter, you will begin to build confidence and an eye for using the digital tools in Photopea. You will also gain an eye for making sure things are aligned correctly and well balanced throughout the composition.

Using guides and rulers and snapping images into place are all key to creating successful designs and printed projects, and taking the professional quality of your work to the next level.

Ruler

The ruler tool acts as a digital tape measure that can find and align precise calculations between objects, lines, and shapes. It can also help you straighten out an image, measure angles, and add a professional quality to your images and designs. To access the ruler guide, go to **View** and find **Ruler** about halfway down the menu. See *Figure 2.16*:

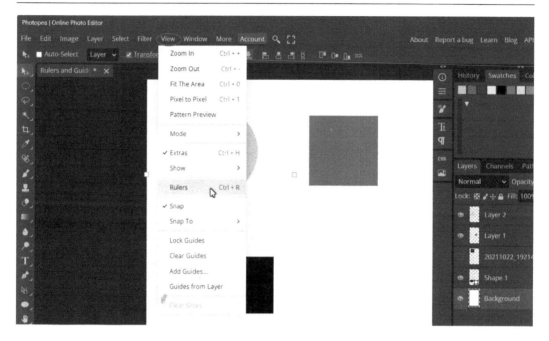

Figure 2.16 – Accessing the Ruler option

The numbered ruler guides (see *Figure 2.17*) will be displayed vertically on the left side of the document and horizontally just above the document.

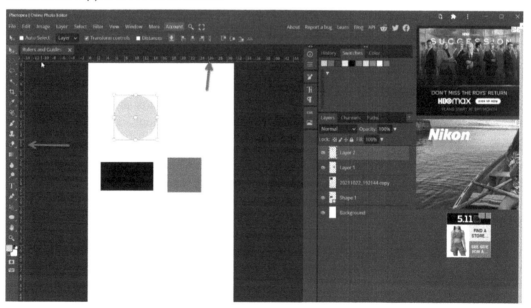

Figure 2.17 – The ruler guides displayed

To hide the rulers, you can go back to **View** menu and select **Ruler** again. You can also use the *Ctrl + R* shortcut to show and hide the rulers.

Guides

Guides are displayed horizontally and vertically over the document to help designers arrange elements more accurately. You can also use the Move tool (see *Figure 2.18*) to remove a guide by dragging the guide back into the ruler.

Figure 2.18 – Move tool

To access the guides and display or hide them, go to **View | Show | Guides**, as in *Figure 2.19*:

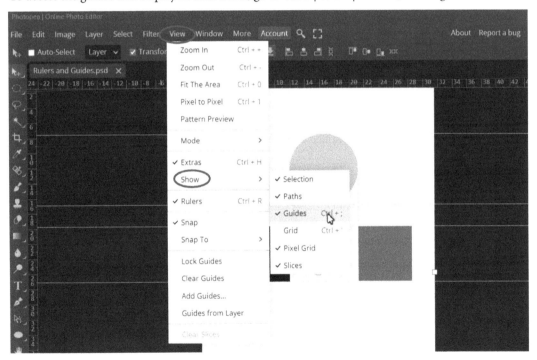

Figure 2.19 – Accessing Guides

You can also use the Move tool (see *Figure 2.18*), to remove a guide by dragging the guide back into the ruler.

Figure 2.20 – Move tool

That covers the overview of accessing guides and how to manage them. Let's move on to the next section.

Using grids

To access grids, go to **View** | **Show** | **Grids**. Grids make it easier for designers to place and arrange images and other design elements on a document more accurately. Grids help to create an even distribution of objects in a document.

Snapping

Using the Snapping feature helps designers and users place objects more precisely with other objects arranged on the document. When you move one object within 5 pixels of another, Photopea will automatically align them for you. This makes it easier to work with fewer mistakes.

Now that we've covered guides, rulers, grids, and snapping, let's move on to the next section where we learn about **artboards** and how to utilize them.

Utilizing artboards

Artboards allow you to have multiple documents within a single (folder) document. This can be useful for organizing images for your social media content, elements, images, and designs for websites, storyboards, comic books, and more.

Each artboard is rectangular and can be rearranged, resized, or swapped with other artboards. Web and app builder programs such as Sketch, XF, and Figma also use artboards. They are used for brainstorming ideas, especially for mapping out the plans to build a website, showing how it will be used and organized.

The following screenshot shows an example of using artboards across different social media platforms for my multimedia project, *Remembering Flash*:

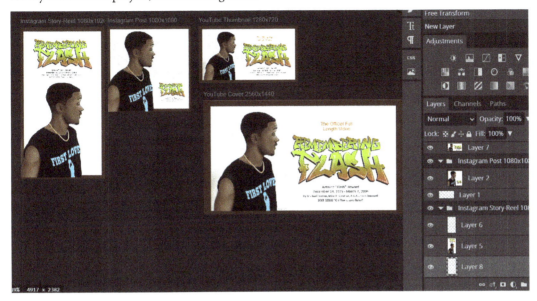

Figure 2.21 – Artboards used for social media

To access artboards, go to the top left toolbar, click and hold the **Move tool** arrow, and move across to **Artboard Tool** in the submenu that appears.

Once the artboard tool is selected, create a rectangular shape in the document with it. A new artboard folder will automatically be created in the **Layers** panel. You will also notice four circles with a plus sign around the four planes of the artboard you created. See *Figure 2.22*:

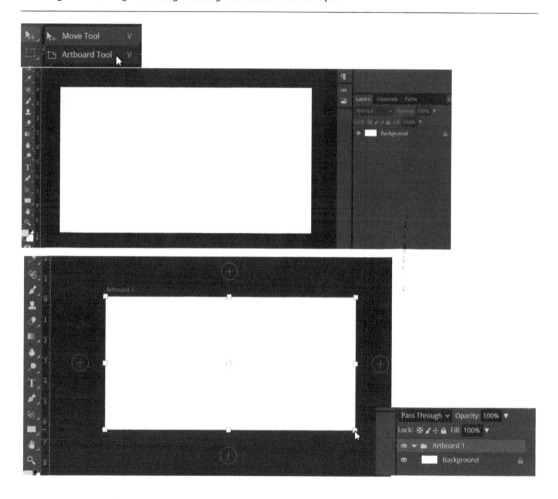

Figure 2.22 – How to select the Artboard Tool and setting artboards up

This was just to give a quick overview of artboards. We will look into artboards later in the book in more detail and do a quick step-by-step example.

This wraps up the basics of understanding artboards and how to set them up.

Summary

In this chapter, we looked at different ways to open a document, save documents, and set up document sizes. We reviewed templates with preset sizes and saw how to set up rulers, guides, and snapping, which enable us to align images and elements into a balanced composition. We also learned the basics of setting up documents for high-quality printing, as well as for viewing documents on the web.

Lastly, we gained a solid understanding of artboards and how they are used for managing multiple pages within a single document.

In the next chapter, we will cover the fundamentals of navigating and utilizing the workspace in Photopea.

3

Navigating and Utilizing the Workspace

In this chapter, we will go over navigating the workspace, explaining the purpose and features of the toolbar of the workspace, using the top menu bar, and using flyout menus to execute your photo editing and designs sufficiently. Don't get intimidated by all the tools, icons, and menus in Photopea's workspace. The more that you work in Photopea, the easier it becomes to remember the tools, prompts, and effects. By the end of this chapter, you'll have a better understanding of what the tools are and how to properly use them for a specific task.

In this chapter, we will cover the following topics:

- Understanding the workspace
- Understanding the top menu
- Understanding the toolbar
- Understanding the sidebar

Understanding the workspace

Once you create a new document or open an existing document after launching the Photopea application, you will see the user interface. The interface has a similar setup to most image editing programs. It can be broken down into four sections to help make it a little easier to navigate and follow along.

See *Figure 3.1* for Photopea's interface breakdown:

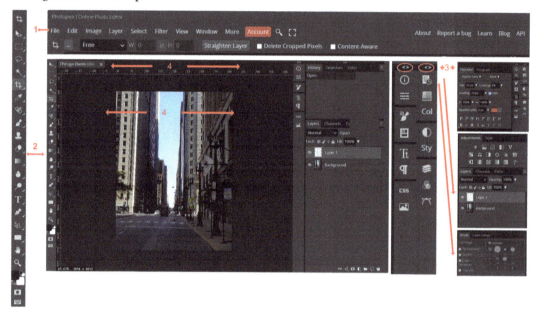

Figure 3.1 – Workspace

Now, let's navigate the four sections of the workspace:

- **1 Top menu**: This is located at the top of the user interface.

  2 Toolbar: This is located on the left side of the user interface. It enables you to access editing tools such as the **Brush** Tool, **Eraser**, **Clone** tool, **Pen** tool, **Dodge** tool, **Burn** tool, and **Spot Healing Brush** tool, to name a few.

- **3 Sidebar**: This is located on the right side of the user interface and consists of tabs you can access such as the **Layers** panel, **Channels**, **Paths**, **Brush Properties**, **Colors**, **Swatches**, and **Adjustments**, to name a few.

- **4 Working area**: This is located at the center of the user interface and is where you will be creating new documents, saving, and editing your images.

Let's dive a little further into the **File** menu located at the top. The **File** menu consists of the following essential functions:

- **New…**: This is for creating a new document

- **Open…**: This is for opening a document

- **Save**: This is for saving a document

- **Print…**: This is for printing a document

- **Export as**: This is for exporting a document

- **Automate**: This is for recording and reusing an action on multiple documents

- **Script…**: This is for creating a script

You can view the following image for further details:

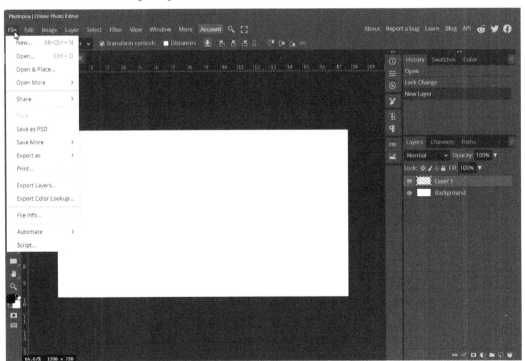

Figure 3.2 – The File menu

Additionally, the top menu includes **Edit**, **Image**, **Layer**, **Select**, **Filter**, **View**, **Window**, **More**, and **Account**. We will become more familiar with the top menu throughout the book:

- The **Edit** menu gives you the power to edit and alter the image in a number of ways. For example, it allows you to undo or redo a change to images and design elements on your document, copy and paste objects and design elements from one layer or document to another, scale objects up and down, align objects, and other features that we will cover in more detail as we execute projects throughout the book. See *Figure 3.3*:

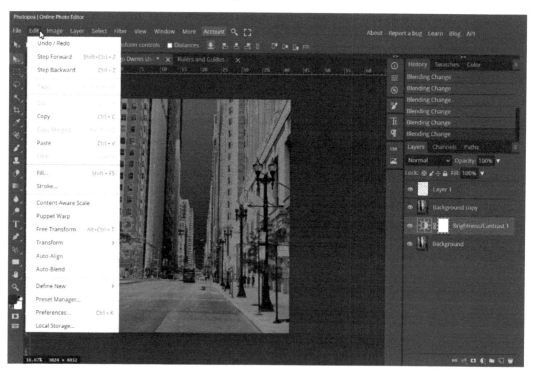

Figure 3.3 – The Edit menu

- The **Image** menu consists of a variety of ways to make adjustments and enhancements to your images and documents if you are willing to take the time to explore and experiment with them. Notice how some options such as **Adjustments** have hidden features, such as **Brightness/Contrast…**, **Hue/Saturation…**, and so on. See *Figure 3.4*:

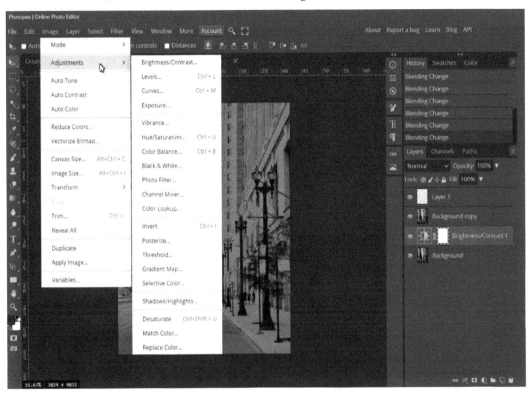

Figure 3.4 – The Image menu

- The **Layer** menu gives you the **Duplicate Layer**, **Duplicate into…**, **Delete**, **Layer Style** (within which you get **Blending Options**, **Inner Glow**, **Stroke**, **Inner Shadow**, **Gradient Overlay**, and other effects), **New Fill Layer**, **New Adjustment Layer**, **Raster Mask**, **Vector Mask**, **Cropping Mask**, **Smart Object**, **Group Layers**, **Arrange**, **Combine Shapes**, **Animation**, **Merge Down**, **Flatten Image**, and **Defringe** options. See *Figure 3.5*:

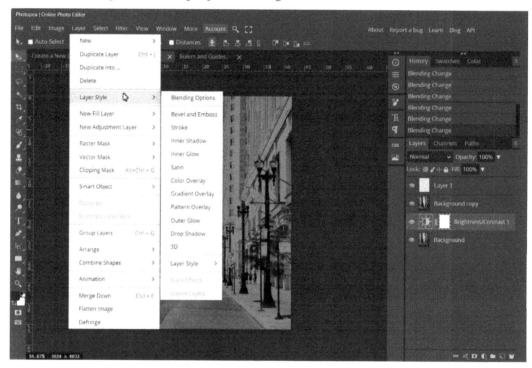

Figure 3.5 – The Layer menu

- The **Select** menu has nice features for things such as refining edges, removing the background, color range, and other things listed under it. See *Figure 3.6*:

Figure 3.6 – The Select menu

- The **Filter** menu contains a variety of **filters**. Filters are digital effects that can quickly apply to your images and selections to create artistic styles, effects, repairs, and overall enhance them with a unique and fresh look. You have a ton of options for adding filters to your images and design elements. 3D, blur effects, distortion, sharpening, and so on, are just a few of thousands of possibilities. See the *before and after effects* of when the **Oil Paint… Filter** is applied to the buildings:

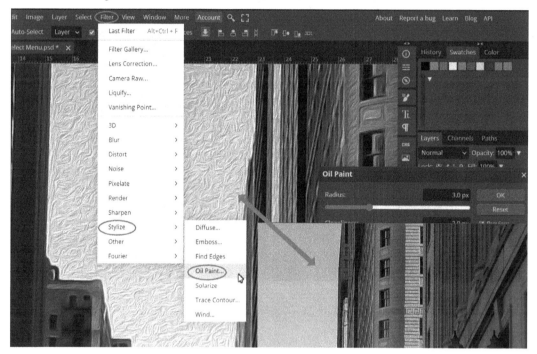

Figure 3.7 – The Filter menu

- The **View** menu helps you with placing, arranging, and aligning your images, text, and design elements within your composition as well as the document in a professional and balanced manner. See *Figure 3.8*:

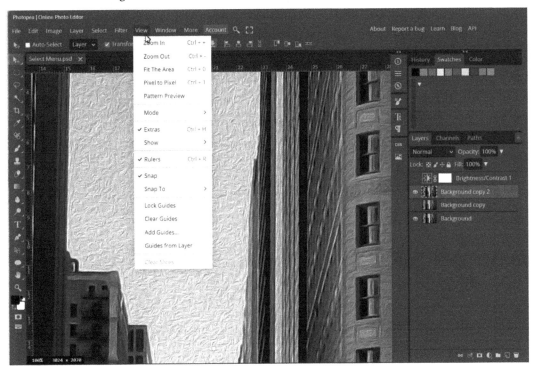

Figure 3.8 – The View menu

- The **View** menu gives you access to hide and show the panels and tabs. For example, you can hide and show your **Layers**, **Color**, and **Swatches** panels, to name a few. See *Figure 3.9*:

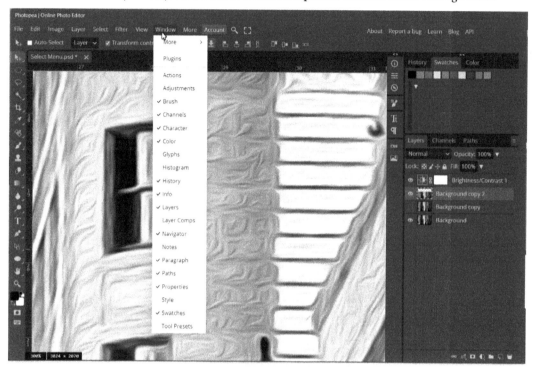

Figure 3.9 – The Window menu

- The **More** menu is for learning more about Photopea, troubleshooting support, reporting a bug, accessing a blog that posts the latest updates and news, and learning where to find the Photopea community on social media. See *Figure 3.10*:

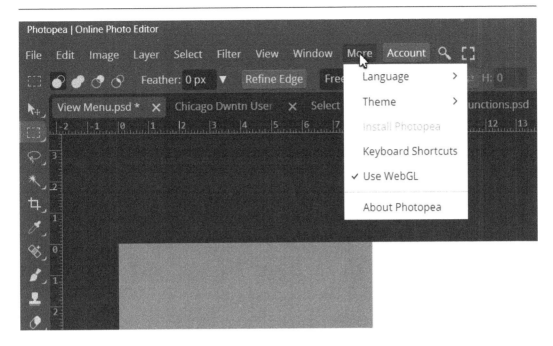

Figure 3.10 – The More menu

- The **Account** menu gives you login options with your Google email, Facebook, GitHub, or Microsoft account. You also have the option to use the Photopea software for free with ads (there is an extension update on Google Chrome to remove display ads for the **FREE** version) or choose the **PREMIUM** account for $5 monthly with no ads, quick email support, and keep twice the number of steps in your **History** tab. See *Figure 3.11*:

Figure 3.11 – The Account menu

- Just below the main menu consists of the tool Parameters area; which displays the information for the current tool being used. Each tool has an option to alter and edit the properties in the Parameter area. For example, the **Rectangle Select** tool gives you options to **Feather** the edges on an object currently selected, or an option to use **Refine Edge** on an object currently selected, and so on. This information will change each time that you select a different tool located on the left. See *Figure 3.12*:

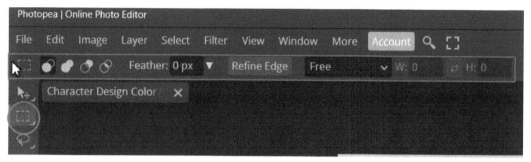

Figure 3.12 – Parameter bar

Now, we are ready to break down the tools in the next section, which will enable us to learn and apply them to execute photo editing more efficiently.

Understanding the toolbar

The toolbar is located on the left side of the interface. It contains all the necessary tools needed to edit your images and design elements on the document.

> **Important note**
> You can only use one tool at a time and will be changing between them often.

When you use the mouse to left-click and hold over a tool, a sub-tool menu flies out with other tools for additional features to assist you. See *Figure 3.13* for the **Move Tool** example:

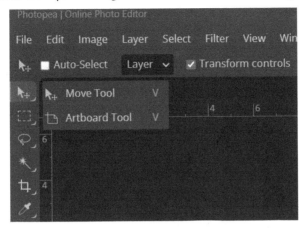

Figure 3.13 – Move Tool

Now that we've introduced the toolbar and the sub-tool menu, let's deep dive into the tools.

There are a lot of powerful tools in Photopea, much like other photo editing programs. We will only cover them briefly in this section. We will provide a list of functions for the main tools at the end of the book.

- **Move Tool** is the first tool located in the upper left corner, just below the top menu. It is used to move, scale, and rotate an object or image on the canvas. When you click and hold the **Move Tool**, **Artboard Tool** is revealed and used for creating multiple documents within a folder (see *Figure 3.13* with **Move Tool** example).

- The **Rectangle Select** is located just below the **Move Tool**, and is used for making selections around images, and objects. This will allow you to move, copy, delete, and or add effects only to the area of the image selected. It has an **Eclipse Select** tool hidden underneath the **Rectangle Select** tool and is used just the same but within the area of the ellipse. See *Figure 3.14*:

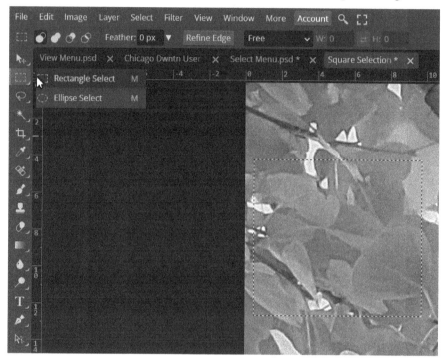

Figure 3.14 – Rectangle Select

- The **Lasso Select** tool is another way you can make a selection, along with the **Polygonal Lasso Select**, and **Magnetic Lasso Select** tools (see *Figure 3.15*); we'll cover these tools in *Chapter 5, Understanding Selection Fundamentals*.

Figure 3.15 – The Lasso Select tool

- The **Magic Wand** tool (see *Figure 3.16*) is also a specialty selection tool that enables you to select the same colors (or very close within the color range) when you click on a spot on your image. Both the **Quick Selection** and **Object Selection** tools are hidden underneath the wand tool. The **Quick Selection** tool is used as a brush tool to select areas by drawing or painting them in. The **Object Selection** tool uses a rectangle that picks up color within the small cross mark (we will cover this in *Chapter 5, Understanding Selection Fundamentals*):

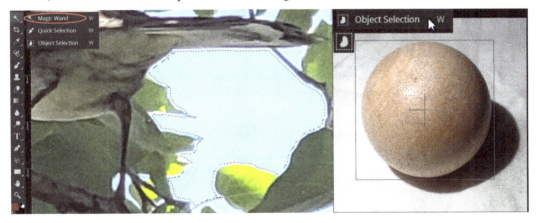

Figure 3.16 – The Magic Wand and Object Selection tools

- **Crop Tool**, as seen in *Figure 3.17*, is used to crop the document or adjust for a different composition, which, in turn, changes the document size and pixels. You can also find the **Perspective Crop**, **Slice Tool**, and **Slice Select Tool** options hidden under it.

Figure 3.17 – Crop Tool

- The **Eyedropper** tool, as seen in *Figure 3.18*, is used to sample any colors from your image, which can be saved as swatches, and keep consistency throughout an image or composition.

Figure 3.18 – The Eyedropper tool

- **Spot Healing Brush Tool**, as seen in *Figure 3.19*, enables you to remove objects, blemishes, and anything else you don't want, by drawing or painting over the area you are trying to edit, similar to the Brush tool. The area will be filled with another area of the image. **Spot Healing Brush Tool** does the same as the Clone tool, except it differs in how it can adapt and unify the surroundings with the same color and area.

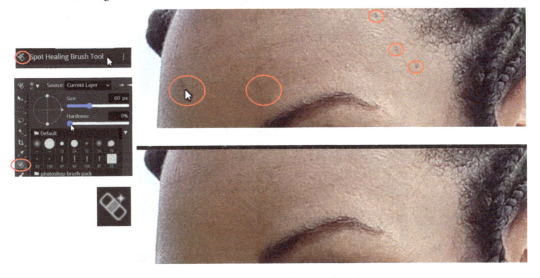

Figure 3.19 – Spot Healing Brush Tool

- **Magic Replace Tool** is a new special effects AI tool. You select it like the **Lasso** tool, but you get a prompt bar to type in some keywords, and with what you have typed, it will find and place a stock image in the selection area for you. (The selection area has to be smaller 1000 x 1000 px for the AI generator to display properly.) You can change the descriptive keywords as often as you like until you are satisfied with the results! (This is a game changer!) See *Figure 3.20*:

Figure 3.20 – Magic Replace Tool

- **Patch Tool** can be executed by first making a section over a specific area you would like to replace. You then drag the selection to a different area of the document. See *Figure 3.21*:

Figure 3.21 – Patch Tool

- The **Content-Aware Move Tool** is another new featured tool that enables you to move an object from one area to another once the object is cut out. The initial area will be filled in with the surrounding background to unify the space. See *Figure 3.22*:

Figure 3.22 – Content-Aware Move Tool

- **Red Eye Tool** is a specific brush tool that allows you to remove the red eyes in a portrait created from a camera flash or secondary light source.

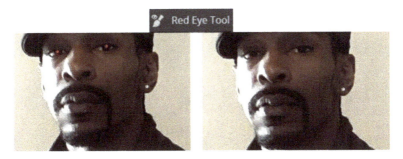

Figure 3.23 – Red Eye Tool

- **Brush Tool**, as seen in *Figure 3.24*, has a vast library for drawing and painting strokes on a layer to create and/or manipulate your images, and edit pixels freely on the layer, or within a specific area you make a selection with. There are tools in the drop-down menu under it, such as the **Pencil** Tool **Color Replacement** Tool, and so on.

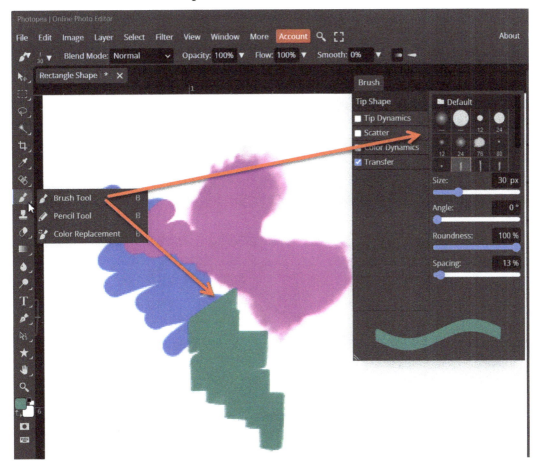

Figure 3.24 – Brush Tool

- **Clone Tool** enables you to clone a certain area of an image on a rasterized layer. See *Figure 3.25*:

Figure 3.25 – Clone Tool

- **Eraser Tool** allows you to erase a part of, or all of, an image's pixels. See *Figure 3.26*:

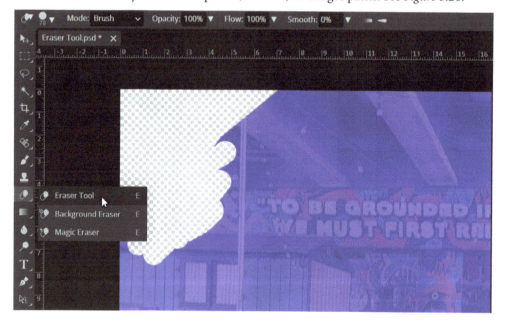

Figure 3.26 – Eraser Tool

- **Gradient Tool** and **Paint Bucket Tool** are used to fill colors in shapes, objects, images, and an entire layer/background. See *Figure 3.27*:

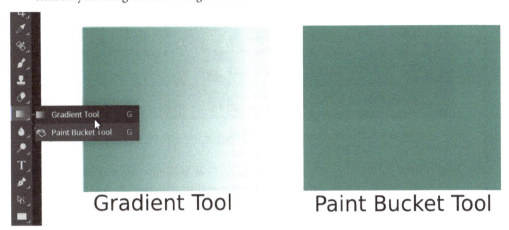

Figure 3.27 – Gradient Tool and Paint Bucket Tool

- **Blur Tool** and **Sharpen Tool** enable you to blur or sharpen areas of an image or element. See *Figure 3.28*:

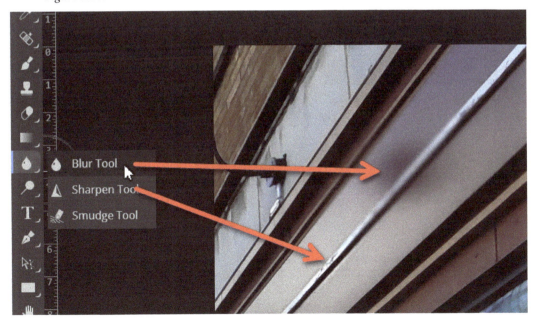

Figure 3.28 – Blur Tool

- **Dodge Tool**, **Burn Tool**, and **Smudge Tool** are great tools often overlooked. **Dodge Tool** can lighten colors, **Burn Tool** will darken colors, and **Smudge Tool** can spread colors around and stretch objects (elongate) or shorten them (squeeze). See *Figure 3.29*:

Figure 3.29 – The Dodge Tool and Smudge Tool sub-menu

- **Type Tool** is used for typing on layers. Adding text can add interest and context to your standalone images, graphics, posters, logos, designs, etc; which can help communicate your brand or message to the viewer.

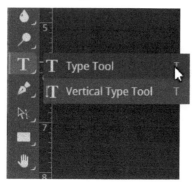 

Figure 3.30 – Type Tool

- The **Pen** tool is a great tool for creating paths, tracing an image to convert it into vector art, and making precise selections. See *Figure 3.31*:

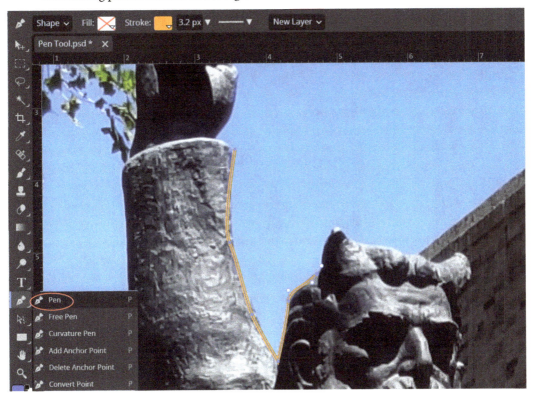

Figure 3.31 – The Pen tool

- The **Path Select** tool allows you to change the shape of an object or change multiple objects by combining them and destroying parts of the shape, or shapes, to make a unique shape.

Figure 3.32 – The Path Select tool

- On the other hand, the **Direct Select** tool allows you to change the shape of an object that is created with paths from either the **Pen** tool or shape tool, and have the blue anchor dots that are visible while you're connecting them or selecting the object. You can manipulate the stroke size (the outline) curve or straighten a path point. You will see the property options change when you select either of the tools in the parameter bar. See *Figure 3.33*:

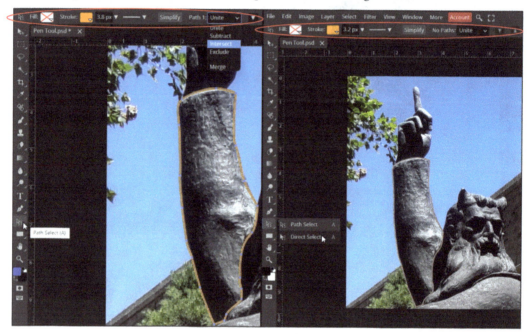

Figure 3.33 – The Path Select and Direct Select tools

- The **Rectangle** shape tool creates squares and rectangles and represents the geometric shapes hidden behind the rectangle icon. The **Ellipse** shape tool creates circles and elliptical shapes at any degree.

- The **Parametric** shape tool can draw polygons, stars, and spirals, which can also be customized further by editing the parameters. Each shape is vector based and can be resized and reshaped with other tools and effects. The custom shape allows you to import custom shapes from other sources if they are saved as CSH files. See *Figure 3.34*:

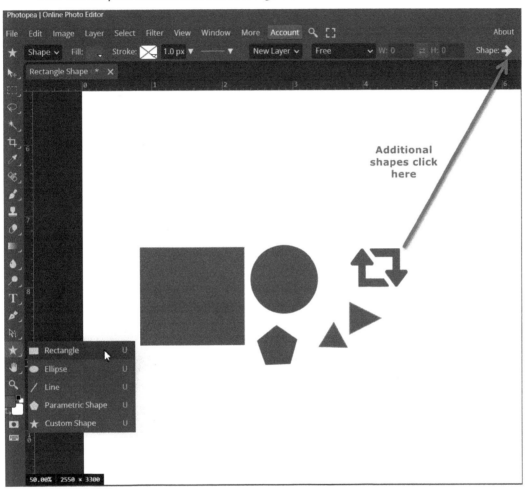

Figure 3.34 – Creating shapes

- Besides creating shapes with the shape tool, you can create lines of different thicknesses, lengths, and colors with the **Line** tool.

- The **Line** tool creates lines at any angle and width. See *Figure 3.35*:

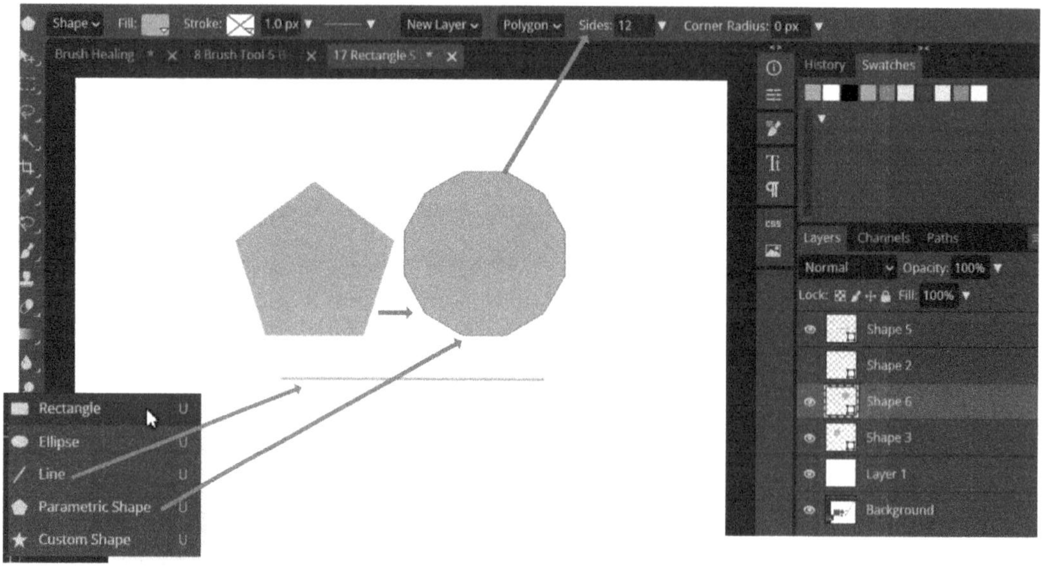

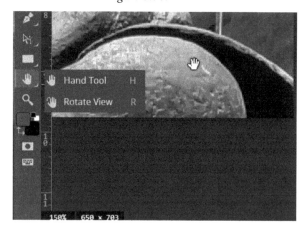

Figure 3.35 – Creating Line shapes and Parametric Shape shapes

- **Hand Tool** enables you to freely pan (or move) to specific areas of an open document by selecting the hand, and then clicking and dragging to the desired area. Clicking the space bar will temporarily activate the tool. The hidden **Rotate View** tool enables you to rotate the entire document around the screen. See *Figure 3.36*:

Figure 3.36 – Hand Tool

- **Zoom Tool** enables you to zoom in and out of the entire document as well as specific areas of the document when you drag the tool to the specific spot you would like to look at. You can also use the shortcut keys, *Ctrl* + space bar, to do the same thing. See *Figure 3.37*:

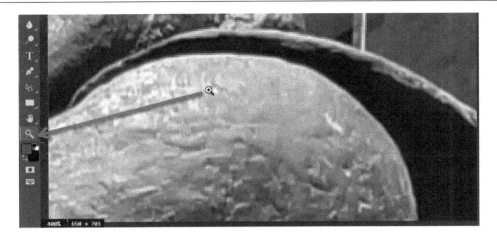

Figure 3.37 – Zoom Tool

The foreground and background color options are located at the bottom of the toolbar. These options let you quickly swap which color will be the main or current color to paint or fill on an object or page. You can also double-click on either square to activate the color picker to change the colors. See *Figure 3.38*:

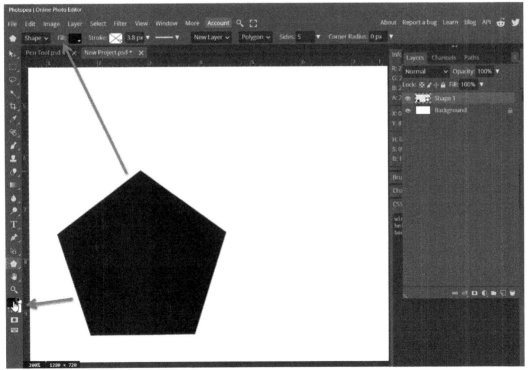

Figure 3.38 – Foreground and background color

That covers the section on the toolbar. Now, we can proceed to the *Understanding the sidebar* section.

Understanding the sidebar

The sidebar is located on the right-hand side of the interface. This section enables you access to work with Layers, Channels, Swatches, Paths, Navigator, Brushes, and a lot of other tools.

The small arrows (<>) above the panel allow you to show or hide the tabs for the Layers, Swatches, and so on, when you left-click on the <>. See *Figure 3.39*:

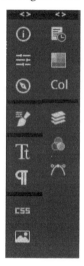

Figure 3.39 – Sidebar

Now, we are ready to cover the fourth section – the main area.

Exploring the main area

The main area displays the document, or multiple documents, currently open (see *Figure 3.40*). You can also rearrange the documents and open or close them from the main area.

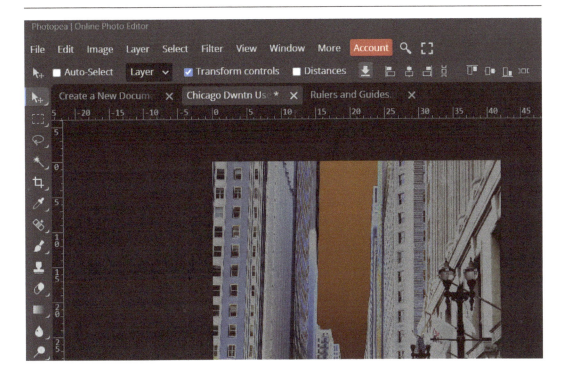

Figure 3.40 – The main area with multiple documents open

To close, in exploring the workspace, we gained an understanding of the five areas of the workspace, can now locate the tools, as well as the sub-tools hidden underneath the tools, and know how to open and save documents in the main area.

Summary

Photopea has a workspace and navigation system similar to other popular photo editing software. We gave an introduction to the top menu, sidebar, and main area, and covered most of the tools in the toolbar to give you a better understanding of the projects we will cover throughout the book.

You won't learn about all of this in one chapter, but gradually, you will apply what we cover and practice, by first carrying out simple tasks before building to more complex tasks as we progress.

In the next chapter, we will learn how to use brushes and how to create custom brushes to suit your needs.

4

Layer Management: Creating, Organizing, and Applying Advanced Features

In this chapter, we will learn about Layers and the various uses and features you can apply to each one. Layers allow us to freely move and edit individual objects within an image or design. This gives us a chance to experiment non-destructively so we know whether the effects or edits will make or break the image or design.

Alongside working freely and experimenting, we can organize Layers by naming each one so we can quickly identify one Layer from another, move a Layer in front or behind another, and group layers into folders. This comes in handy when you have, say, 10, 20, or more Layers to manage. Other Layer features include creating masks, Layer properties, effects, and smart objects.

In this chapter, we will cover the following topics:

- Understanding layers
- Organizing layers
- Applying masks to layers
- Layer styles (also called layer effects)

Understanding layers

Layers are created and maintained in PSD files, better known as Photoshop Documents, found in most image editing programs. Layers allow users to work on individual images, design elements, and text on one layer at a time within the same documents. Layers help in creating an overall design or finished image.

Think of a Layer as a single sheet of tracing paper or a sheet of glass, so that you can draw, paint, write, or add some kind of effect on top of others. For example, you may have a tough time deciding whether a portrait looks better in a specific color. You could easily fill in or paint the background on a separate Layer behind the portrait to quickly see what works. Or, maybe you want to see if you should add a sky or a tree. It's possible to do this by hiding, revealing, and organizing the Layers in Photopea's **Layers** panel (see *Figure 4.1*):

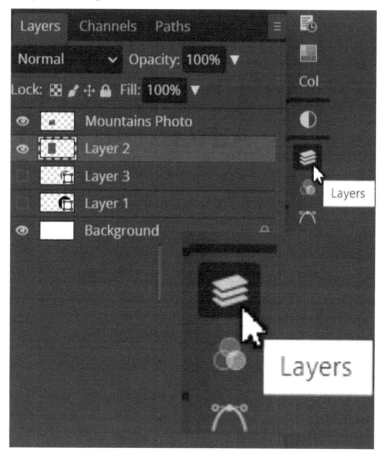

Figure 4.1 – Layers panel

Let's break down the **Layers** panel in the upcoming section.

The Layers panel

The **Layers** panel is located on the sidebar, to the right of the current document or the main work area, used for manipulating, organizing, and building upon Layers. Notice the **Layers** icon resembles three sheets of paper stacked on top of each other to represent the active layers to the left of it. They are labeled **Background**, **Layer 1**, **Layer 2**, **Layer 3**, and **Mountains photo** (the top layer). The bottom layers are at the back, while the top layers are in front.

You only work on one Layer at a time, but you can also select multiple Layers to rotate them, delete them, or group them all into a Layer Folder or a subfolder.

Layer Folders can be useful for the following:

- Organizing a large number of Layers into one Folder
- Creating a folder within a folder that can be opened or closed
- Merging or flattening layers into one Layer.

Now that we understand the **Layers** panel, we can move on to creating Layers in the next section.

Creating layers

Creating Layers allows us the freedom to move, edit, and organize our images and elements non-destructively, as well as to figure out whether design decisions are going in the right direction. Let's begin creating layers in a new document:

1. After you've launched Photopea, create a new **1224 x 792** pixel landscape view document (or your preferred size).
2. Next, go to **File | Open Mountains photo**.
3. Notice the white-colored layer name is **Background**, and it is locked (cannot be edited). Click the lock icon on the layer to unlock it.

4. Next, double-click the name with the mouse pad to rename the Layer as Mountains Photo (this helps keep layers organized). You can also drag **Mountains photo** directly from the documents folder into the new document without having to rename the Layer. See *Figure 4.2*:

Figure 4.2 – Create a new 1224 x 792 pixel document

5. Next, you'll notice the **Mountains photo** layer has a small black square. This means it is a **Smart Object** layer (we'll cover this in *Chapter 15, Tips, Tricks, and Best Practices.*). Simply right-click with the mouse and select **Rasterize Layer** and the small black square will disappear (see *Figure 4.3* and *Figure 4.4*):

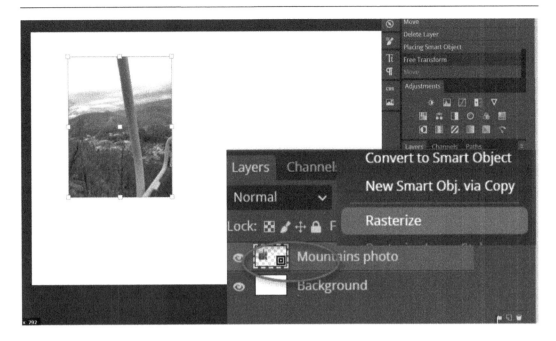

Figure 4.3 – Right-click the Mountains photo Layer and Rasterize the image

The image is officially a **raster layer**, and the black square on the Layer has been removed successfully. See *Figure 4.4*:

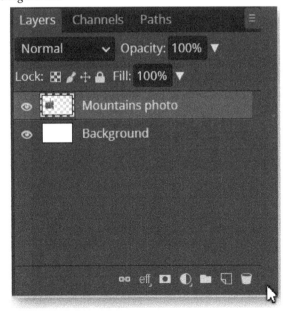

Figure 4.4 – Mountains photo converted from Smart Object to raster layer

Raster Layers are made up of pixels that can be edited destructively and non-destructively with adjustment Layers, filters, brushes, and so on.

6. Next, right-click **Mountains photo** and select **Duplicate Layer** (this allows us to make a copy of the same Layer). This is useful as a backup in case the original gets permanently altered, and so on.

With the duplicate Layer, we can also explore different ways the Layer can behave in comparison to the original and not worry about losing the original for now. See *Figure 4.5*:

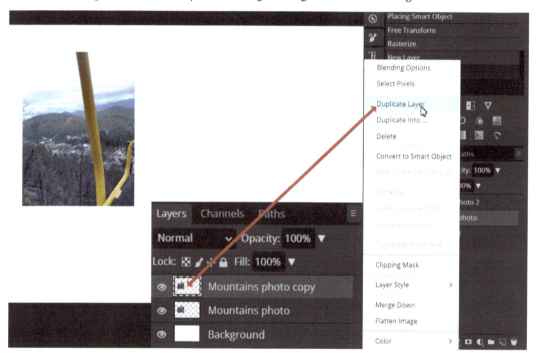

Figure 4.5 – Make a duplicate layer of Mountains photo

Now that we've created a document, rasterized the image, and made a duplicate Layer of **Mountains photo** as a backup in case of a mishap, we are ready to move on to the next section, *Organizing layers*.

Organizing layers

Organizing Layers will allow our workflow to run more smoothly, save time, and make photo editing more enjoyable:

1. Next, I opened the brown rectangle image called **Fig 4.6 brown Rectangle** and place it into the **4.Mountains Border** document as a Layer.

2. Right-click the mouse over the brown rectangle to convert it to a rasterized Layer. You can either go to **Layer Menu**, **Arrange**, **Send to Back**, or just drag the **Fig 4.6 brown Rectangle** layer below **Mountains photo**. See *Figure 4.6*:

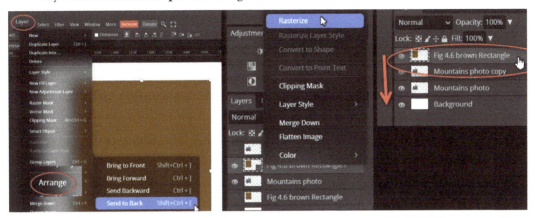

Figure 4.6 – Convert rectangle to raster layer and send to back of Mountains photo

The brown rectangle acts as a border around **Mountains photo** (see *Figure 4.7*):

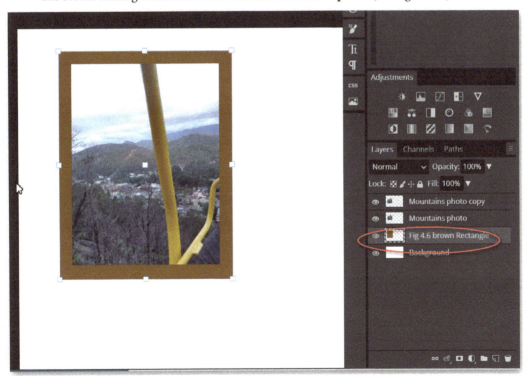

Figure 4.7 – Adding a rectangle border around Mountains photo

We have learned different ways to arrange Layers in order and have learned how to make a photo more interesting by adding a border. That sums up this section. Now we are ready to review *Group layers* in the next section.

Group layers

Group Layers will come in handy when you have a document that uses a high number of Layers. It would get overwhelming trying to work with, and maintain a smooth workflow, with 50 loose Layers. Grouping and naming certain effects and elements together will help immensely. Let's take a look at the next exercise:

1. Select both **Mountains photo** layers, then click **New Folder** at the bottom of the **Layer** panel to group them inside it.

    The photos will be hidden inside the **New Folder**. You can click on the Folder's small arrow to reveal them.

2. Next, double-click the folder and rename it **Mountains photos**.

    Naming your Layers and Folders helps organize and locate things quickly, when you have a lot of Layers to keep track of (see *Figure 4.8*):

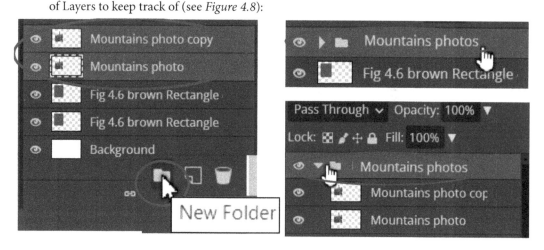

Figure 4.8 – Group layers into a folder

That sums up how to group Layers into a folder. We will practice this further in more complex exercises. Let's begin the next exercise, *Deleting layers*.

Deleting layers

As you continue working on and experimenting with layers, you'll find deleting Layers is useful to get rid of Layers you no longer need and is also a form of organizing and managing layers. There are three ways to delete a layer. Let's take a look at them.

Firstly, you can right-click on the Layer and select **Delete**. See *Figure 4.9*:

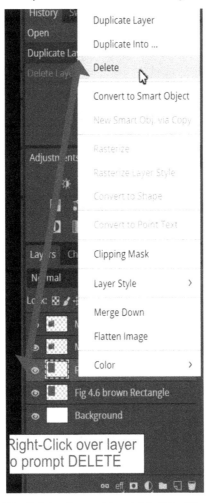

Figure 4.9 – Delete a layer with a right-click of the mouse

Secondly, select the Layer to delete and hover over the trash can located at the bottom right of the **Layers** panel. See *Figure 4.10*:

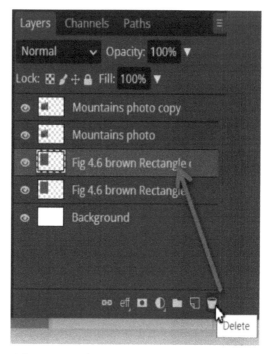

Figure 4.10 – Delete a layer using the trash can icon

Lastly, select the **Layer** menu at the top and click the **Delete** button (see *Figure 4.11*):

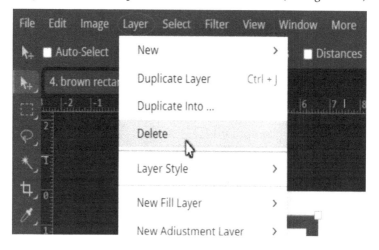

Figure 4.11 – Delete a layer from the Layer menu

After successfully covering deleting Layers, we can move on to the *Locking and hiding Layers* section.

Locking and hiding layers

The most common method to lock Layers is by selecting the padlock Icon. It's located just above the top Layer, next to **Fill: 100%**. Locking Layers protects the Layers from being altered or deleted unless you unlock them. See *Figure 4.12*.

> **Important note**
> If the padlock icon is hidden, click the burger menu above the Layers and select **Lock** to make it visible.

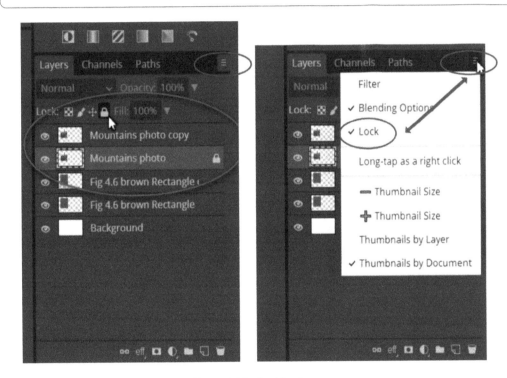

Figure 4.12 – Locking Layers

To hide objects on a Layer, simply click on the small eye icon on the left side of the Layer. Click the eye again to make the Layer visible again. See *Figure 4.13*:

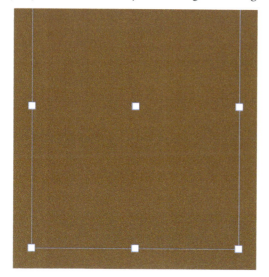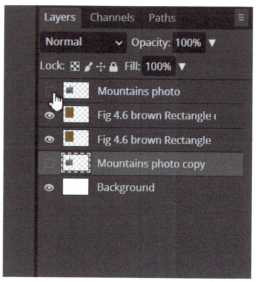

Figure 4.13 – Hide layers

Now that we've learned how to lock and hide Layers, let's look at arranging Layers.

Arranging layers

The **Arrange** options allow you to re-arrange Layers located in the top **Layer** menu. You have the **Bring to Front**, **Bring Forward**, **Send Backward**, and **Send to Back** options.

You can also bring Layers to the front and back by selecting and dragging a Layer above or below another Layer. See *Figure 4.14*:

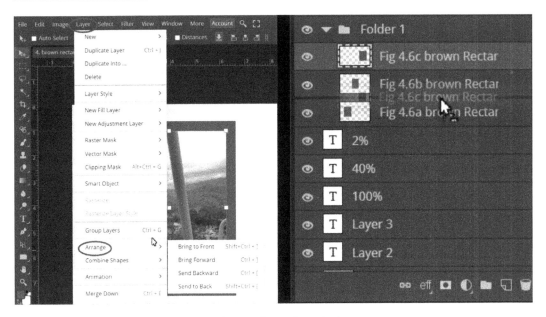

Figure 4.14 – Arrange layers from the Layer menu

We can now move on to Folders and Layers in the next section.

Folders and layers

Photopea and other image editing programs similar to Photoshop allow you to create folders for storing other Layers and other Folders inside main Folders.

A good example is how you are able to create main Folders and subfolders within the main Folders on your computer:

1. To create a Folder, click the small Folder icon at the bottom of the **Layer** panel.

2. You can also select a Layer(s), right-click the mouse pad, and group Layers into a Folder; or from the **Layer** menu at the top menu bar.

That covers Folders and Layers. Let's move on to Layer opacity in the next section.

Layer opacity

Layer opacity determines how transparent or opaque a Layer is. A layer at **100%** is opaque. The lower the **Opacity** percentage, the more transparent a Layer is, and the less you can see it. I reduced the **Mountains photo Opacity** to **40%**. See *Figure 4.15*:

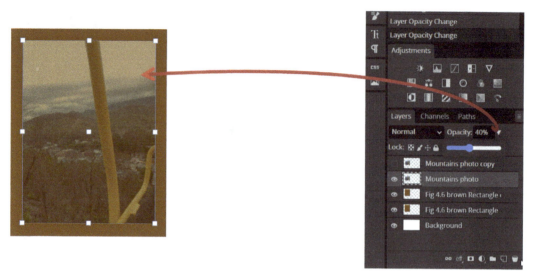

Figure 4.15 – Changing a layer's opacity

In the next example (*Figure 4.16*), we show **Opacity** at **100%**, **40%**, and **5%**, which is nearly transparent. You can type in the layer opacity manually or click the small, white down-arrow button, to slide left or right to change the **Opacity** percentage.

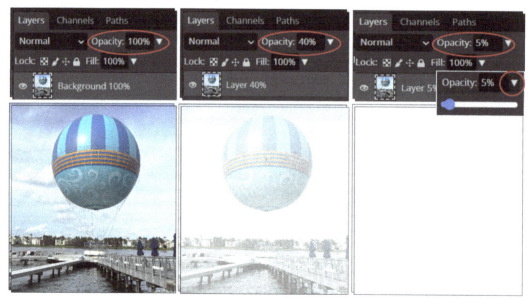

Figure 4.16 – Layer opacity comparisons

That covers layer opacity. Now let's move on to the *Flatten Image* section.

Flatten Image

You can flatten an image to reduce the file size and this also ensures a client can view it in an email or web browser. **Flatten Image** is also useful for sending a document to a printer that doesn't support layer files, reducing the file size and the number of layers to manage.

You can combine all your Layers into one. Here's how to do that:

- Flatten all of the images into one image by *right-clicking* on the Layer and choosing **Flatten Image**. You can also go to the **Layer** menu at the top left and scroll down to **Flatten Image**. Use this with caution, as you cannot revert an image back to the original Layers unless you *undo* changes before you close the document. See *Figure 4.17*:

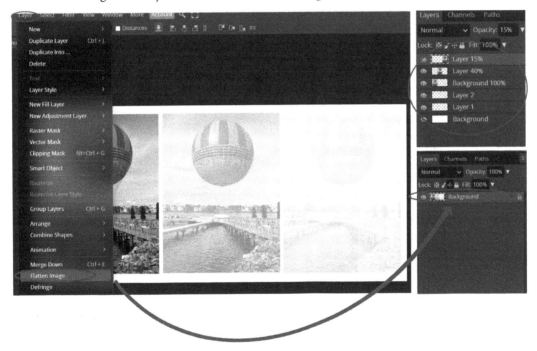

Figure 4.17 – Flatten Image

Now that we've covered **Flatten Image**, let's look at **Merge Layers**, and how they differ.

Merge Layers

Merging a Layer or a number of selected Layers is similar to using **Flatten Image**. The key difference is you only merge (flatten) a specific Layer or a number of selected Layers, whereas **Flatten Image** will combine all of the Layers.

In the example, we merged three Layers into one Layer or image.

To combine the three Layers into one, without the background layer, do the following:

1. Select the top Layer, hold *Shift* while dragging the mouse over the Layers below to the last image, and *left-click* the mouse to select all the Layers.

2. Flatten all three images into one image by *right-clicking* on the Layer and choosing **Merge Layers**. Use this with caution as you cannot revert the image to the original Layers unless you *undo* changes before you close the document. See *Figure 4.18*:

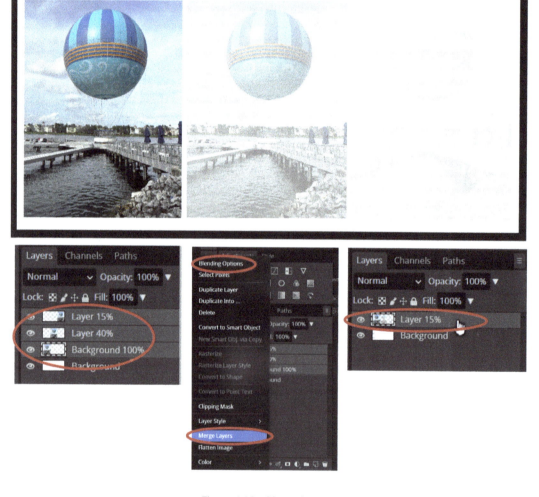

Figure 4.18 – Merge Layers

Now that we understand the difference between **Flatten Image** and **Merge Layers**, and how to apply these functions, we can jump into the next section.

Applying masks to layers

Layer Masks enable you to hide parts of an image or layer non-destructively without erasing the original area or object permanently. They are excellent tools for cutting out objects, removing backgrounds, modifying colors and backgrounds, and controlling the layer opacity.

Layers behave similarly to masks, with limitations. Masks allow us more flexibility to work non-destructively on a specific area on a Layer, whereas Layers will apply opacity and other modifications to the entire Layer.

Let's explore masks and learn how to use them in the next exercise.

> **Important note**
> I recommend using a drawing tablet for painting and erasing out areas of a Mask, especially when you need to mask out complex objects and images.

Photopea has two types of Masks: **Raster Mask** and **Vector Mask**, each with a specific purpose.

A raster Mask is a bitmap-based image that can be edited with the paintbrush and selection tools and uses resolution for quality input. A raster Mask (like the Photoshop Layer Mask) allows you to work on specific layers non-destructively without removing or altering the pixel information. Masks also work within the parameters of the Layer you are working on underneath it.

A Mask can be created on a Layer(s) and inside of a folder(s). When you are working on a Mask, you will only see two colors – white and black. White will reveal what is on the Layer, while black will hide all or part of a Layer, depending on how much of the area you need to cover.

Both raster and vector Masks are linked to a Layer by default. It's indicated on the Layer as a link (similar to a chain). This enables you to move them around together with the **Move** tool. You can also unlink masks so you can move them around independently of one another.

Vector masks are created with the **Pen** or **Shape** tool, and just like vector graphics, have unlimited scaling capability that is not reliant on bitmap resolution. Vector masks can be useful for placing an image on different shapes, such as a circle, square, star, or custom shapes, and have several other functions. You'll see an example of this in *Figure 4.21* in the upcoming section.

You can create a raster Mask with the following steps:

1. Select the Layer you would like to add a raster Mask to and select **Add Raster Mask** located at the bottom of the **Layers** tab. Or, you can go to the top **Layer** menu, scroll down to **Raster Mask**, and select **Add (Reveal All)** to create one.

2. Notice the Mask icon next to the Layer is white. This means everything on the Layer is visible, and *foreground color black* will hide pixel data when you apply black with the brush (see *Figure 4.19*):

Figure 4.19 – Creating a raster mask

3.   Next, select **Brush Tool** and increase the brush size to cover large areas. Make sure the *foreground color* is *black* (located in the lower left of the toolbar).

4.   Begin painting with black and the area will become transparent (hidden).

      If you select the color white and paint over the same area, it will reveal what was hidden. See *Figure 4.20*:

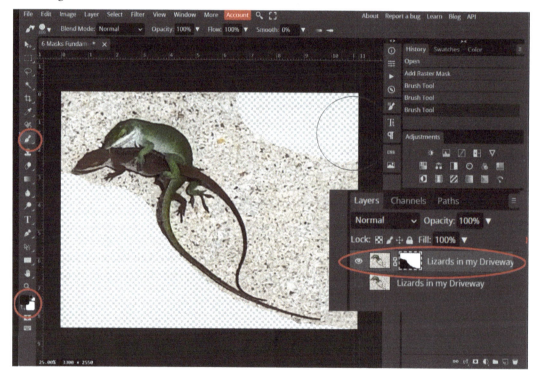

Figure 4.20 – Applying the mask with Brush Tool

You can also *double-click* on **Raster Mask** to prompt options such as **Density** and **Feather** (to soften the edges of the selection) and revert the mask to select outside of the shape rather than within the shape. See *Figure 4.21*:

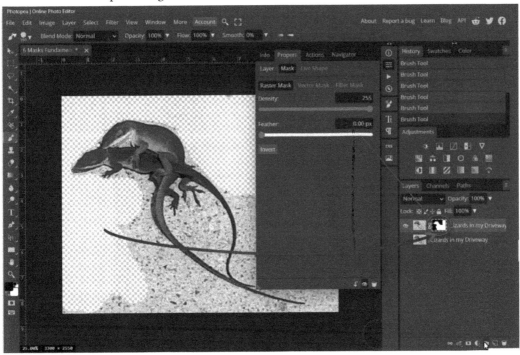

Figure 4.21 – Double-click the mask to prompt more property modifications

Now that we've learned how to create a mask to hide and reveal an area of a Layer, we can move on to the next section, *Creating a vector mask*.

Creating a vector mask

Vector masks can turn specific areas of an image transparent while revealing the other areas inside a shape, text, and so on. You can also make more precise selections with the **Pen** tool to make a vector mask, but it requires more time and effort. We will review this in more depth in *Chapter 5, Understanding Selection Fundamentals.*

Let's follow along to make a vector mask using a Polygon from the **Shape** tool:

1.   Create a shape with the **Shape** tool. Drag it below the image/photo (see *Figure 4.22*).

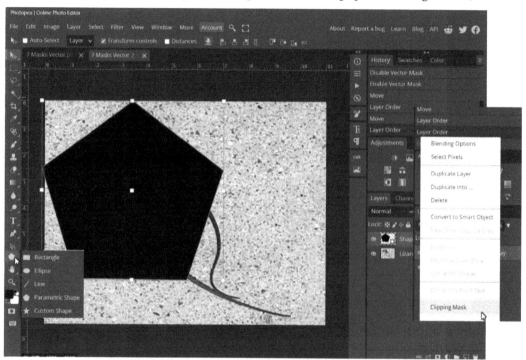

Figure 4.22 – Creating a clipping mask with the Shape tool

2.   Next, right-click the photo or image and select **Clipping Mask**. You should see the image cropped into the shape that was below it. See *Figure 4.23*:

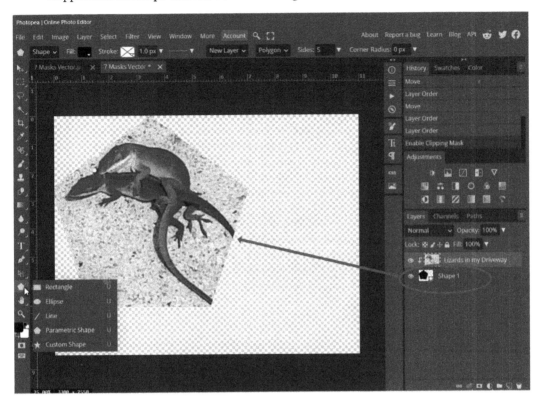

Figure 4.23 – Image cropped within a shape using a clipping mask

3. With the **Move** tool, you can also resize and adjust the image within the Mask as desired. See *Figure 4.24*:

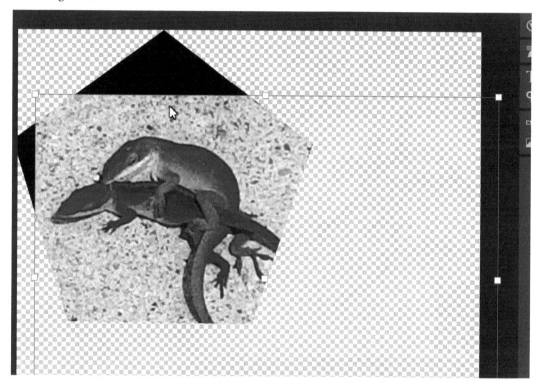

Figure 4.24 – Adjust and resize the image with the Move tool

You can access the **Raster Mask**, **Vector Mask**, and **Clipping Mask** prompts from the **Layer** menu (see *Figure 4.25*):

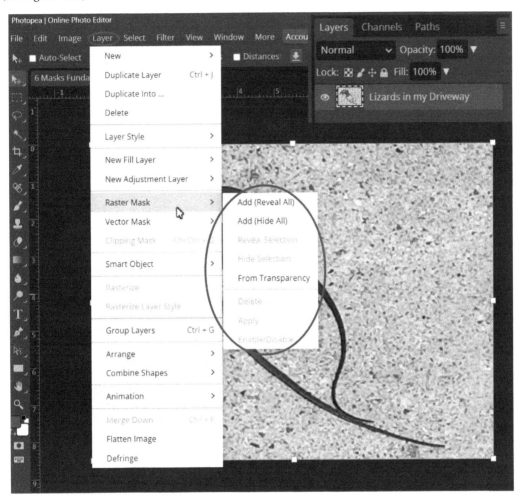

Figure 4.25 – Raster Mask, Vector Mask, and Clipping Mask in the Layer menu

We have learned how to create clipping Masks with the **Shape** tool and access Masks in the Layer menu. Let's move on to the next section, on Layer styles.

Layer styles (also called layer effects)

Layer styles are used to create powerful effects that can be applied to images, photos, text, design elements, and shapes very quickly and non-destructively. Some examples of layer effects include **Bevel and Emboss**, **Stroke** (outline), **Drop Shadow**, **Inner Glow**, **3D**, and so on. Each effect has its own properties that can be adjusted accordingly when it is selected in the **Blending Options** panel.

Double-click on a Layer to access the **Layer Style** panel.

You can apply more than one effect by selecting the effects you like with a checkmark. Just know that you need to select each one individually to make an adjustment to that specific effect. See *Figure 4.26*:

Figure 4.26 – The Layer Style panel

Let's take a look at some examples of Layer Styles used in the examples.

The **Stroke** and **Drop Shadow** layer effects have been applied to the border of the swan image. See *Figure 4.27*:

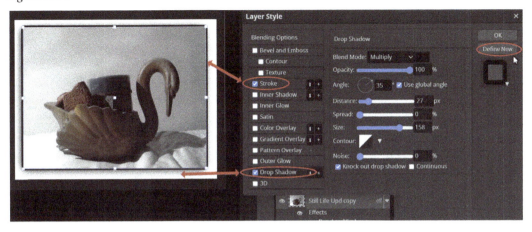

Figure 4.27 – Applying a border using layer effects

In the next example, we apply Layer Styles to the text.

Double-click on the Layer titled **LAYER STYLES** to open the **Layer Style** window.

Next, *right-click* on the layer to open **Blending Options**, for **Stroke**, **Inner Shadow**, **Pattern Overlay**, **Drop Shadow**, and so on. See *Figure 4.28*:

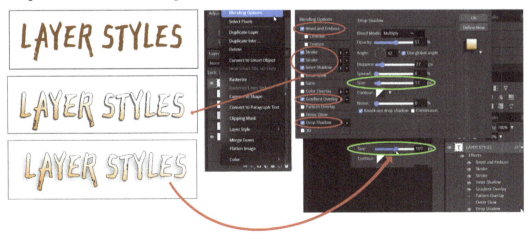

Figure 4.28 – Apply layer styles to text

You can also click on the **Effects** Layer below the image Layer and hide or reveal effects whenever you'd like to. Also, whichever blending option you click on, the **Properties** window will open for that specific one. For example, if I click on **Drop Shadow** under **Blending Options**, the **Drop Shadow** properties window will appear. See *Figure 4.29*:

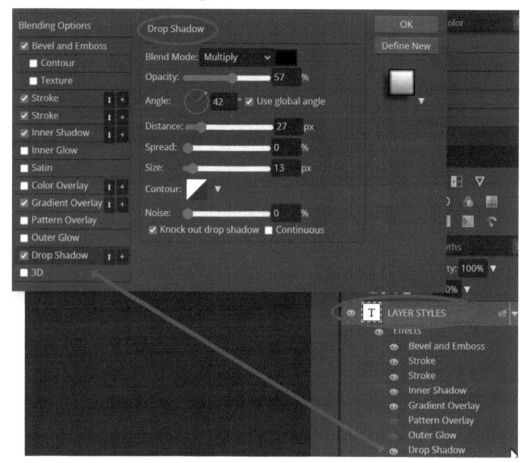

Figure 4.29 – Hide, reveal, and layer styles on layers, and open specific blending options

We will use Layer effects in other projects throughout the book. In the meantime, don't be afraid to explore and experiment with the effects on different images, shapes, and text.

New Adjustment Layer

Using adjustment Layers gives you more versatility and adjustability for image edits using non-destructive adjustments.

You can turn your images to grayscale (black and white) and adjust colors, hue, and saturation without altering the pixels on the original image.

Additionally, adjustment Layers have their own Masks, opacity, blend modes, and so on.

To make a new adjustment Layer, follow along:

1. Click on the small circle icon at the bottom of the Layer, as seen in *Figure 4.30*:

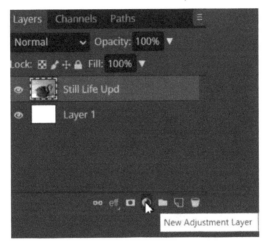

Figure 4.30 – Make a new adjustment layer

2. Next, select a **Vibrance** adjustment for this example. Dragging **Vibrance** and **Saturation** all the way to the left, change the image to grayscale. Notice the adjustments are applied above the original image Layer. See *Figure 4.31*:

Figure 4.31 – Changing an image to black and white using adjustment layers

The adjustment Layer features are found under the **Layer** menu located in the main menu at the top.

In the third swan still life example in *Figure 4.32*, we turned it to a monochromatic green color by applying a number of non-destructive adjustment Layers using **Hue** and **Saturation**. We can adjust and fine-tune them, as well as hiding and deleting them. See *Figure 4.32*:

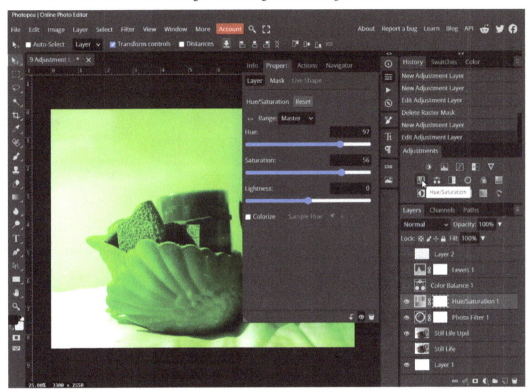

Figure 4.32 – Changing color, hue, and saturation with layer adjustments

That sums up the essentials of applying adjustment Layers.

Summary

In this chapter, we covered a lot of ground, learning about Layers and working within a document efficiently.

We learned how to access Layers, name and organize Layers, group Layers into Folders and subfolders, edit Layer properties, and create Masks, adjustment Layers, Layer Styles, and Layer properties, and also learned about other features related to the chapter. We have gained a lot of information and skills and unlocked a lot of powerful non-destructive editing tools and features that will prepare us for the next chapter.

In the next chapter, we will explore selection fundamentals and learn how to use the best selection method for certain types of images and objects.

5
Understanding Selection Fundamentals

In this chapter, we will learn about the different selection styles. A **selection** enables us to isolate an area of an image. The selected area will be the only thing affected once a selection is established. We are able to make selections on raster layers containing pixels. By reflecting on *Chapter 1, Taking Your Design and Editing to the Next Level with Photopea,* we know that pixels are tiny individual rectangular colored dots. The more pixels you have, the better the quality and more details you can add to an image or digital painting, and the easier it will be to make selections.

By the end of this chapter, you will be able to identify and apply the fundamental techniques of making various selections on different objects, figures, and other elements ranging from simple shapes with color, to more complex selections such as human hair.

In this chapter, we will cover the following topics:

- Making selections
- More on the Magic Wand tool
- Advanced selection techniques
- Refining edges
- Working with channels

Making selections

Photopea has a variety of selection tools to assist in making selections on all objects, ranging from simple solid shapes with flat hard edges, to more complex images of animals, trees, and even humans with hair that would require refining the edges to select.

The selection tools consist of the **Magic Wand**, **Quick Selection**, **Lasso**, **Polygonal Lasso**, **Magnetic Lasso**, **Rectangle Select**, **Elliptical Select**, **Object Selection**, the **Pen tool**, and the newly developed **Magic Replace AI tool**. Overall, we will use the selection tools to select, move, replace, and remove an object or background from an image, and use non-destructive masks to make layer adjustments and enhancements with filters, color correction, sharpening, and more.

Let's take a look at an example of an object that has been removed or separated from the background with the **Quick Selection** tool. See *Figure 5.1*:

Figure 5.1 – Quick Selection tool example

We will cover how to remove the sphere from the background in the *Applying the Selection Tool* section.

Now that we have been introduced to the different tools used to make selections, let's get started with putting the selection tools into practice in the next section.

Putting the selection tools into practice

We could have easily used the Magic Wand tool for all of the geometric shapes in the examples *Figure 5.2 - Figure 5.5*, but instead, we will try out a variety of other selection tools including: the Magic Wand, Rectangle Select, and Elliptical Select tools, to gain an understanding and sense of how to use them.

> **Important note**
> You will know when an object or area is selected by the tiny marching ants moving around the edges.

Let's take the steps to select a *Polygon* using the **Magic Wand** tool as follows:

1. Go to **File | New | New Project** | select **Print| Letter Size 8.5 x 11**.

2. Go to **File| Save as PSD** and rename the file to something like **Selections Practice**.

3. Create a *New Layer* and rename it *Polygon*.

4. Select the **Parametric Shape** from the **Shape** tool located on the lower left side of the **Toolbar**.

5. *Double-click* on the **Fill** button and change the *Polygon* color to red.

6. Next, *select* the **Magic Wand** and make a selection on the red polygon for a quick demonstration. The marching ants should be moving around the edges of the polygon. See *Figure 5.2*:

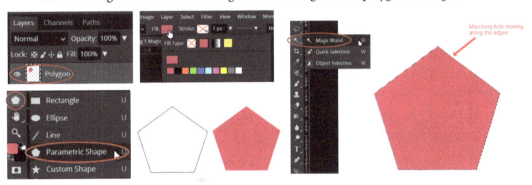

Figure 5.2 – Select the red polygon with the Magic Wand tool

Now that we've tested the **Magic Wand** tool on a *Polygon* using the **Shape** tool, let's jump into the next section to experiment further with creating shapes and making selections.

Creating shapes and making selections

Now that we have a little practice making a Polygon with the Shape tool and a selection with the Magic Wand tool; let's dive further into different Selections methods, and create a square, circle, and triangle using the Shape tool.

Creating a triangle requires an extra step with the Polygon. Let's cover it in the following steps:

1. Continue working on the same document **Selections Practice** from the previous section.

2. Create a *New Layer* and rename it *Triangle*.

3. Select the **Parametric** tool in the **Toolbar**, and create a **Parametric Shape, Polygon**; but *type in the number 3* for the **Sides** to create a triangle. The **Sides** box appears when you create a **Parametric Shape**, and is located just below the top menu, to the right of the **Account** button. See *Figure 5.3*:

Figure 5.3 – Make a New Layer and Create a Triangle with the Parametric Shape tool

4. Now that we've made the triangle, create a **New Layer** and rename it *Square*.

5. Next, select the **Rectangle** tool and create a purple color square.

6. Create a **New Layer** and rename it *Circle*.

7. Next, select the **Ellipse** tool and create a green circle to the right of the purple square. You should have something similar to what I have. See *Figure 5.4*:

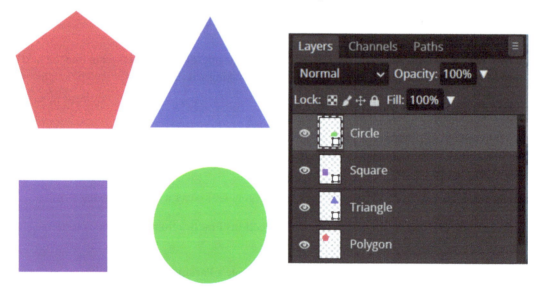

Figure 5.4 – Create a Square and Circle with the Shape tool

Now that we created our shapes, we can practice making selections and move them around. Let's move on to the next section.

Making selections with the Rectangle and Ellipse Select tool

We will start our next Selection exercise with the purple square.

1. Select the **Rectangle Select** tool and click on the *Square Layer* with the mouse to make sure it's the active Layer.

2. With the **Rectangle Select** tool active, make a selection around the purple square, and a bounding box should appear around the selection.

3. You can make adjustments to the selection by switching to the **Move** tool and adjusting it with the blue anchor points until the marching ants are as close to the edges as possible. See *Figure 5.5*:

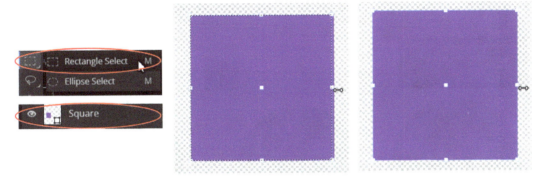

Figure 5.5 – Rectangle Select tool and the bounding box

Important note

You can Press *Ctrl + D* to deselect the selection (a shortcut), around the purple square; otherwise, you can click **Deselect** under the **Selection** menu located at the top of the interface.

Now that we've made a selection using the Rectangle Select tool, let's move on to the next selection exercise.

Making selections with the Ellipse Select tool

In this exercise, we will look at the **Ellipse Select** tool and also learn how to correct issues that may occur when making selections in this next series of steps.

1. Make sure the green *Circle Layer* is active and select the **Ellipse Select** tool located in the **Rectangle Select** tool as a sub tool tab.

2. Make a selection around the green circle.

3. The green circle wasn't fully selected with the **Elliptical Select** tool. Once I switched to the **Move** tool to move the selected circle, it got cropped leaving only the part of the circle that was selected.

You can fix this issue in a number of ways:

- Press *Ctrl + Z* or (*Command + Z*) on Mac to *undo* the changes.

- Try to make a more accurate **selection** with the selection tool again by making sure you draw completely around the circle so that it doesn't get clipped.

- Or, while the **selection** is *active* with the marching ants displayed, you can slightly hover the mouse over the marching ants to show a small white cross. This allows you to move the active selection itself, without affecting the green circle.

- You can *right-click* the active selection and select **Transform Selection**.

  You will see a *double-sided arrow* appear. Use it to adjust the selection to fit around the circle. See *Figure 5.6*:

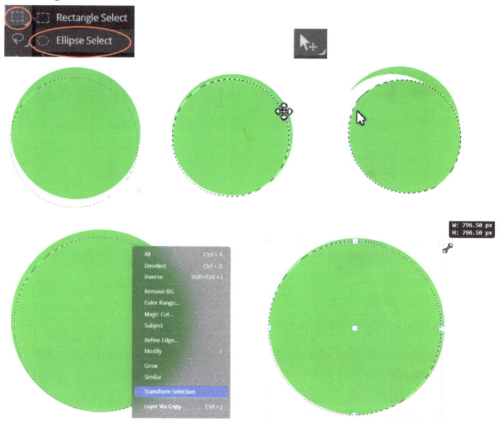

Figure 5.6 – Green circle not fully selected with the Elliptical Select tool

> **Important note**
>
> Again, using these selection tools was just for practice to get familiar with using them, and develop an understanding of the issues you may face with selections and how to fix them.

After learning of the issues that can occur like the green circle in *Figure 5.6*, another problem that can arise is moving shapes around on the same Layer. In this example, I flattened all of the shape Layers into a single Layer. Once I selected the blue triangle, I moved it on top of the purple square, forgetting that the shapes are not on separate Layers.

Once I committed to the placement of the triangle, it merged with the purple square. While the document is still open, I can *Undo* the move by pressing *Ctrl + Z*.

If I would have saved the changes and closed the document, the blue triangle would have been permanently merged with the purple square. See *Figure 5.7*:

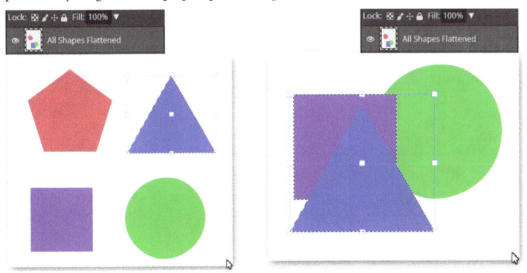

Figure 5.7 – Moving selected objects into a different area

> **Important note**
>
> Be careful and intentional when moving the selected objects in different areas of the document. You can accidentally drag or paste a shape or element onto the same layer of another object, merging them together with undesired results.

That concludes the introductory section on making selections, and correcting issues that can occur working with them.

Let's move on to the *More on the Magic Wand tool* section.

More on the Magic Wand tool

I will use the **Magic Wand** tool on the following image of the blue hooded man (see *Figure 5.6*) to give you a better understanding of how selections work in conjunction with the **Feather** and **Tolerance** parameters. (Usually, the **Quick Selection** tool would be recommended here instead for speed):

1. Grab the Blue hooded photo reference located in the `Chapter 5 Resources` folder and drag it into the Photopea Main area.

2. Select the Magic Wand tool, and set **Feather** to **11** and **Tolerance** to **30**. Start with a medium brush size to select the hoodie. Notice it only selects part of the hoodie.

3. Hold the *Shift* key down and continue to expand the selection, or hold down the *Alt* key while clicking the wand to decrease the selection near the selection area that went outside of the hoodie. This will get the selection as close as possible on the hoodie, and off of the grey background.

4. For smaller areas, such as the face, I changed **Feather** to **5 px** and **Tolerance** to **25 px**. This gives the Magic Wand tool more accurate selection ability with smaller areas. I continued holding the *Shift* key and clicking the Magic Wand until it covered the entire figure. See *Figure 5.8*:

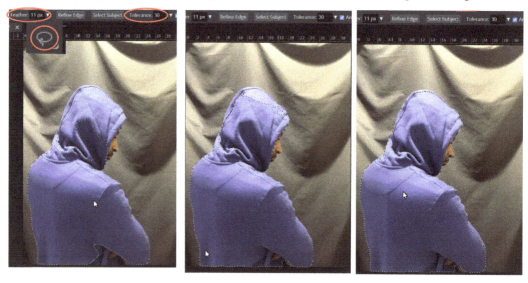

Figure 5.8 – Using the Magic Wand with the Feather and Tolerance parameters

5. Next, go to the **Select** menu and choose **Inverse** to reverse the selection and make it cover the space outside of the figure. This allows you to remove or adjust the background if desired. See *Figure 5.9*:

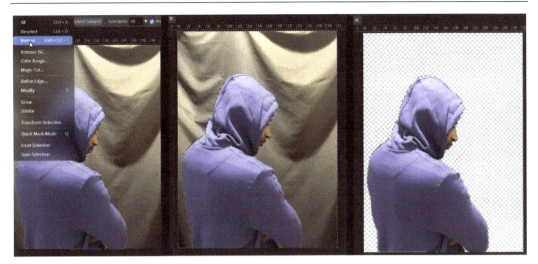

Figure 5.9 – Inverse the selection and create a raster mask

6. For now, we will leave the image as is. We will revisit this image later for an advanced exercise in the *Working with Channels* section.

> **Important note**
>
> You may have to gradually decrease the brush size to make detailed selections on small areas. Also, when a selection is active, you can expand the selection by holding the *Shift* key while left-clicking the mouse, and decrease the selection area by holding the *Alt* key down while clicking the left mouse button.

That covers the Magic Wand tool, let's move on to the next section.

Applying the Quick Selection tool

The **Quick Selection** tool is great for finding and selecting the edges of an object. We can see this in full effect on the sphere example in *Figure 5.8*. Here's how it's done:

1. Grab the Sphere photo reference located in the Chapter 5 Resources folder and drag it into the Photopea Main area.
2. Duplicate the Sphere Layer as a backup copy and hide it.
3. Select the **Quick Selection** tool located under the **Magic Wand** tool.
4. Click the mouse on the sphere area and it will find the hard edges for you automatically.
5. Once the sphere is selected, you can edit it, enhance it, create a mask, and so on. In this case, let's cut and paste it from the background to make it a separate layer.

6. To separate the sphere from the background, press *Ctrl + X*.

7. The sphere will disappear until you press the *Ctrl + V* to paste it onto the document as a separate Layer from the background. See *Figure 5.10*:

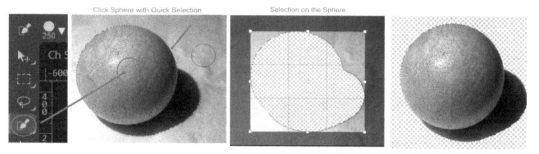

Figure 5.10 – Select the sphere with the Quick Selection tool

Before you **Save** the changes to the sphere document, let's experiment further with this selection.

8. Undo the changes up to the **Quick Selection** tool's first selection of the sphere.

9. Go to the **Select** menu and select **Inverse** selection on the sphere. Now we can remove the cloth background, which is the negative space surrounding the sphere.

10. Press the *Delete* key. Now the cloth background has been removed, leaving a transparent background behind the sphere. See *Figure 5.11*:

Figure 5.11 – Inverse the selection and create a raster mask

Now that we understand how to use the **Quick Selection** tool and the **Inverse Selection** technique, let's move on to the next selection exercise.

The Object Selection tool

The **Object Selection** tool operates slightly differently from the Magic Wand. Let's follow the steps on using it.

1. Select the **Object Selection** tool.

2. Pull or drag it around an area of an image or object; in this case the sphere.

3. A small cross appears in the middle of a square; this is where the selection starts and expands outward, selecting the same color (or similar) within the rectangle area. See *Figure 5.12*:

Figure 5.12 – Using the Object Selection tool on the sphere

4.   Notice how the **Object Selection** tool adjusts to smaller selections and different pixel colors and values. See *Figure 5.13*:

Figure 5.13 – Selecting smaller areas and different values with the Object Selection tool

That wraps up the Quick Select tool. Let's take a look at the Lasso tool in the next section.

The Lasso tool

The **Lasso** tool lets you draw selections loosely by hand (similar to sketching or drawing). In the following example image of a statue, I drew a curved line on the side with the Lasso tool for practice. Notice how the Lasso automatically closes the selection once you release the mouse button. Here's how to do it:

1.   Grab the Statue photo reference located in the `Chapter 5 Resources` folder and drag it into the Photopea Main area.

2.   Choose the **Lasso** tool in the toolbar or press *L* to activate the selection mode. Click, hold, and drag the mouse around the statue head; don't release the mouse or stylus until you have drawn the path from start to finish. If you let go too soon, you will have to redraw the selection again (which is why I prefer using a drawing tablet with the Lasso tool). See *Figure 5.14*:

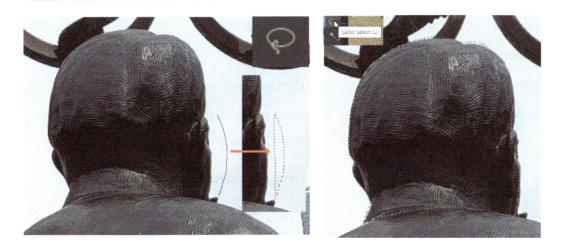

Figure 5.14 – Using and releasing the Lasso tool

3. Once you've made your selection, you'll notice the Lasso isn't perfect. You can modify it with the **Magic Wand** tool so it's closer to the edges of the intended selection. Make sure to adjust the **Tolerance** and **Feather** parameters if the **Magic Wand** is selecting too much or not enough. I had to lower the **Tolerance** value to pick up small areas without going too far over the edges. See *Figure 5.15*:

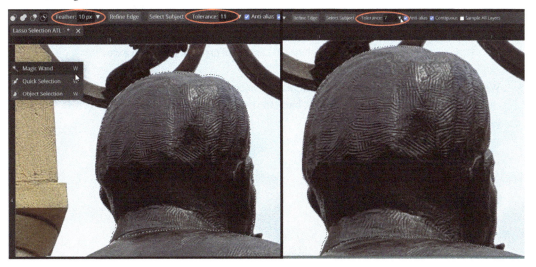

Figure 5.15 – Adjusting the Feather and Tolerance with the Magic Wand tool

4. We can also use the eraser tool to remove any of the background sky that shows, or mask it out non-destructively if we needed to use it in a different way. That's a different topic, for now, we can leave it at is.

That covers the Lasso tool, so let's move on to the Polygonal Lasso tool.

The Polygonal Lasso tool

The **Polygonal Lasso** tool enables you to draw selections in polygonal shapes.

1. Grab the Excavator photo reference located in the Chapter 5 Resources folder and drag it into the Photopea Main area.

2. Select the **Polygonal Lasso** tool in the toolbar or press *L* to activate it.

3. For this example, make a selection around the excavator. Begin the selection with a single click on the lower-right part of the image.

4. Move the mouse upward along the edge of the excavator, using a second click further along to create the first line to begin the path. Each time you click, another line connects the current point to the previous point, like you're connecting a railroad track. See *Figure 5.16*:

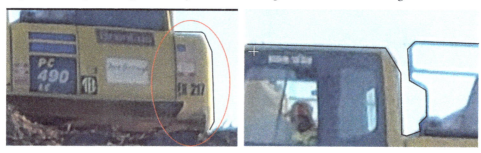

Figure 5.16 – Drawing with the Polygonal Lasso tool on an image of an excavator

5. When you're ready to close the path and make a selection, click on the original point, or double-click anywhere, and the path will lock back to the beginning and create the selection (shown by the marching ants). See *Figure 5.17*:

Figure 5.17 – Connecting the first and last points added with the Polygonal Lasso tool

Normally with a selection like this, you might select and remove the small sections where the sky is showing through (you can do so if you want to practice!), but for now, we just want to get comfortable with the selection tools.

That concludes this section on the Polygonal Lasso tool, so let's move on to the next section.

The Magnetic Lasso tool

The **Magnetic Lasso** tool is similar to the standard Lasso, but with a smart feature that sticks to the edges of an object when making selections.

When you make the first point and move along an edge of an object to draw the selection, the control points will be mathematically calculated automatically as you draw.

It will create the best path from the last control point to where you presently position the mouse. You can continue adjusting the path manually by deleting the last control point; you can also click on the path to add a control point. To finish the selection, double-click the mouse or press *Enter*.

That concludes the Magnetic Lasso tool, let's move on to the following *Advanced selection techniques* section.

Advanced selection techniques

The more projects you work on, the more you will realize there is no one-solve-all selection tool or technique for all of your images and photos (although new tools are always being created to make selections and editing easier). You will have to constantly make minor or major adjustments depending on each image's texture, outline thickness (if it has an outline), edges, lighting, colors, contrast, transparency, and so on.

To be prepared for these situations, let's start by looking at the **Selection Mode** options.

Combining selections with the Selection Mode options

Each selection tool has specific parameters that define how selections behave to assist with your selections. For example, when you select the **Magic Wand** tool, the properties bar for the Magic Wand will appear just below the **File** menu bar. Just to the right of the Magic Wand icon, you will see four circle icons that represent different selection modes. See *Figure 5.18*:

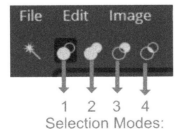

Figure 5.18 – Selection Mode Options: Replace, Unite, Subtract, and Intersect

These modes are as follows:

1. Grab the Cone shape photo reference located in the `Chapter 5 Resources` folder and drag it into the Photopea Main area.

2. **Replace** – The **Replace** selection mode allows you to make a new selection if your initial selection wasn't as accurate

3. **Unite** – Allows you to continue to expand a selection, or combine two selections together.

4. **Subtract** – This removes the first selection, reducing the selection size

5. **Intersect** – This adds an intersecting selection to your initial selection

Now that we understand the how to combine selections and selection modes, let's dive further into selections in the next section.

The Pen tool

The **Pen** tool can be used to make selections with an advantage over all of the other selection tools.

The Lasso, Magic Wand, and others rely on pixels to create temporary selections; once the selection is made, you can't go back and edit the selection if you missed a spot, or over selected an area.

The **Pen** tool creates permanent paths versus temporary selections. The Pen tool allows you to go back and edit selections, regardless of how many steps you made around the object. A lot of users tend to shy away from the Pen tool because it can take a little more time and effort compared to the other selection tools.

> **Important note**
>
> We will use the **Pen** tool for a less complex exercise in this section, *Understanding selection fundamentals*. We will use it in a more complex project in *Chapter 9, Exploring Advanced Image Compositing Techniques,* for a variety of selections.

Now let's get started with the **Pen** tool exercise.

The **Pen** tool is located on the lower half of the toolbar. You create paths with it by clicking to place anchor points one by one. See *Figure 5.19:*

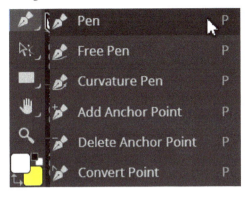

Figure 5.19 – Accessing the Pen tool

Now that we understand how to combine selections and understand the differences and advantages of the Pen tool, let's dive further into using the Pen tool in this exercise.

1. For this example, select the **Pen** tool, and choose **Shape** mode for the pen mode to trace the geometric cone.

2. Open the photo of the cone located in `Chapter 5 Resources` folder.

3. In the Pen tool's properties, set **Fill** to **none**, **Stroke** to a green color, and the stroke thickness to **3px**. (Bear in mind that the stroke thickness will vary with different document DPI resolutions; for example, **3px** on **300 DPI** will look normal, but using **3px** on a **72-DPI** document will be double the width).

4. As shown in *Stage 1* in the following figure, click and release once to create an anchor point, move the mouse down to the lower right and click to place the second anchor point. Then move and drag the mouse to the left, and the antennas will appear, allowing you to create a smooth curve, as shown in *Stage 2. Stage 3* shows what happens when you don't click, hold, and drag. Continue spacing out the placement of your anchor points, as shown in *Stage 4*. See *Figure 5.20:*

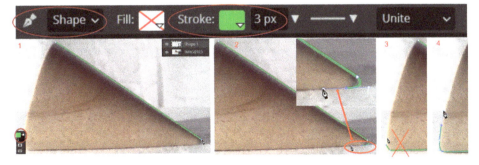

Figure 5.20 – Trace the image with the Pen tool in Shape mode

Important note

If you place consecutive anchor points quickly, they will create straight lines. If you drag the mouse slightly to the left or right, two antennae-like lines will appear. This enables you to change the path from a straight line to a curve. The more you drag the mouse left or right while placing the anchor points, the wider and smoother the curves will be. The anchor points will close into an active selection once you connect the last anchor point to the first one.

5.   Once the selection is made, switch the Pen mode from **Shape** to **Path** and then click on the **Make Selection** button that appears; as shown in *Stage 5 and 6*. See *Figure 5.21*:

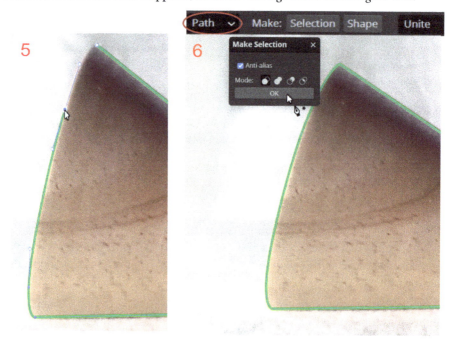

Figure 5.21 – After Tracing the image switch to Path mode

6.   The outline will turn into a selection with a dashed outline.

7.   Go to **Edit | Fill** and choose a color. Change the outline color to **none**. See *Figure 5.22* for the finished result:

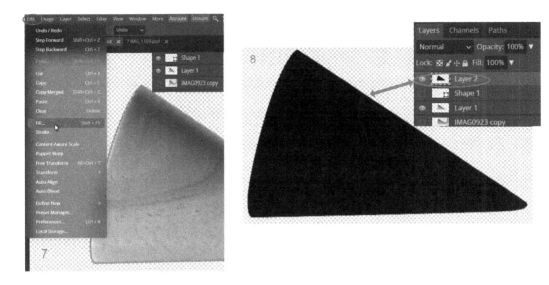

Figure 5.22 – Trace the image with the Pen tool in Shape mode

That covers the basics of using the Pen tool to make a selection, and also apply a **Fill** color to the shape we created from it. Let's go over some of the other features regarding the Pen tool and making selections.

8.   There are several ways to make a selection with the additional **Pen** tools. Click and hold the Pen tool icon until a flyout tab opens to reveal a variety of Pen tools to create paths for selections, vector art, lines, curves, shapes, Add Points to the path you traced, and so on. Let's take a look at them. See *Figure 5.23*:

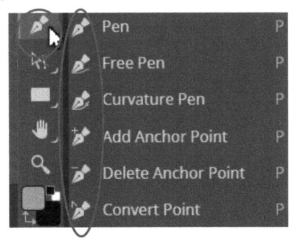

Figure 5.23 – Additional Pen tools

Here is a breakdown of what each **Pen** does:

- **Free Pen** gives you a natural freeform method of drawing, like you would with a physical pen and paper. When you draw, the Pen tool will automatically create anchor points with smooth curves, and fill in the path when you release the mouse button or drawing stylus (if you have a drawing tablet).

- **Curvature Pen** lets you draw smooth, flowing curves with few anchor points. This may save time adjusting the curves manually with the default Pen tool.

- **Add Anchor Point** lets you add extra anchor points anywhere on the path you created.

- **Delete Anchor Point** lets you remove any unwanted anchor points that may be overcrowding a path that needs to be smoother.

- **Convert Point** can edit the anchor points on a path (outline of a shape), as well as vector shape masks. It can convert a smooth anchor point, which is good for creating a curve; or it can create a corner anchor point, which is good for creating straight lines, and corners with angles. For example, 90-degree angles are used to create a square or rectangle. This comes in handy if you don't want to manually convert a line or curve with the pen tool.

To Convert an anchor point, simply position the cursor over an anchor point you would like to edit. If you're editing a smooth point, click it once and release it to create a corner point. When you have a corner point, click and drag away, until the direction line is smooth to your liking.

Changing the Pen Mode

As we continue to learn more about the Pen tool in upcoming chapters, we will run into situations where we need to change the Pen mode for things like, converting your Pen traced path into a selection; similar to the **Magic Wand** and **Quick selection** tool.

The benefit of being able to do this with the **Pen** tool is that we can continue to turn the selection on and off, as well as edit the selection.

To change the **Pen** mode, do as follows:

While the **Pen** tool is active, scroll to the Pen tool properties bar, located in the top left corner, and you can see where you can switch the Pen mode from *Path*, *Shape*, and *Pixels*. We cover this in *Chapter 9, Exploring Advanced Image Compositing Techniques,* in the Man Walking exercise, and throughout the book. See *Figure 5.24:*

Figure 5.24 – Switch the Pen Mode

Now that we've touched on the Pen tool, additional Pen tools, and the Pen mode; you have a better understanding of how to utilize it in your projects, and will know which selection tools to use for various projects throughout the book. Let's explore the next section on refining selection edges.

Refining edges

There are a number of tools and filters we can use to smooth out hard and rough edges of selections we've made non-destructively using masks.

For the first example, we will soften the edges on a banana:

1. Grab the Banana photo reference located in the `Chapter 5 Resources` folder and drag it into the Photopea Main area.

2. Use the **Quick Selection** tool to select the banana. The tool missed a small portion on the end of the banana. To fix this, select the **Magic Wand** tool with the selection still active and press *Shift* while clicking the Magic Wand to expand the selection to cover the entire banana (don't worry about the selection being perfect). See *Figure 5.25:*

Figure 5.25 – Quick Selection on a banana

3.   Next, create a raster mask (using the option located at the bottom of the **Layers** panel) around the banana. Make sure the foreground color is black (located on the bottom of the toolbar) so our mask hides the current background. Also, make sure the mask is active (you should see a dashed white outline around the layer mask). See *Figure 5.26:*

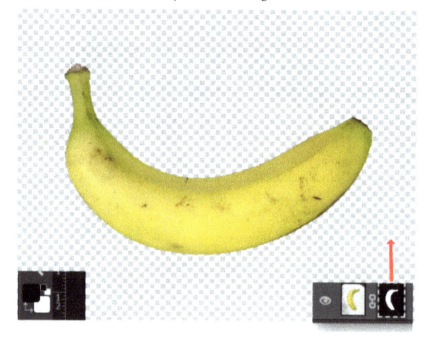

Figure 5.26 – Make foreground color black and create a raster mask

4. Click **New Adjustment Layer** and select **Color Fill** (fill it with black). See *Figure 5.27*:

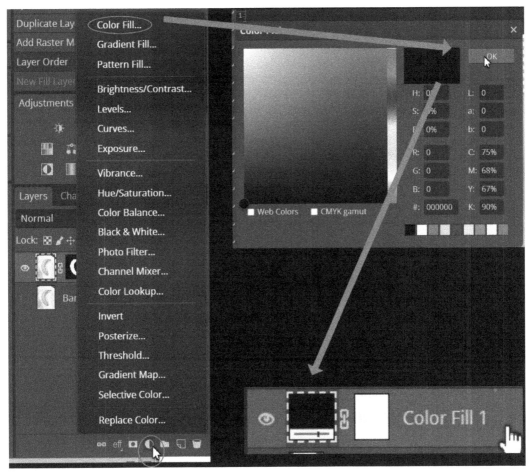

Figure 5.27 – New Adjustment Layer with black Color Fill

5.   In the **Layers** panel, drag the black adjustment layer underneath the banana layer. See *Figure 5.28:*

Figure 5.28 – Drag the black Color Fill layer behind the banana layer

6.   The black background mask reveals how accurate the selection is when we place it behind the banana. You can see where the edges around the banana are rough and jagged. See *Figure 5.29:*

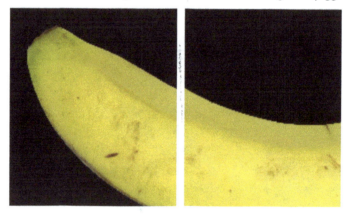

Figure 5.29 – Rough and jagged Lines from the selection

7.   We could smooth this out with the **Refine Edge** tool, but in this case, we will try applying a **Median** filter around the mask layer (not the banana). This option is faster and takes less computing power.

8. Go to **Filter | Noise | Median** to adjust the **Radius** in the **Median** properties bar to about **17px**. See *Figure 5.30:*

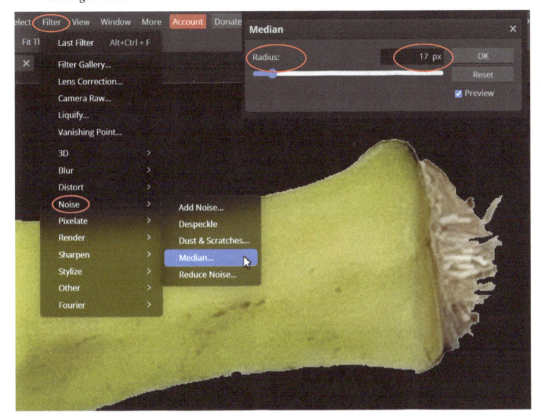

Figure 5.30 – Add a Median filter to smooth the edges

9. Next, with the **Mask** layer still active, go to **Properties** under the **Window** menu (or the **Properties** tab next to the **Layers** tab) and select **Feather**.

10. Increase the **Feather** properties slightly to soften the edges of the banana (just enough to still look like a natural banana). I increased it to **1.00 px** (see *Figure 5.31*):

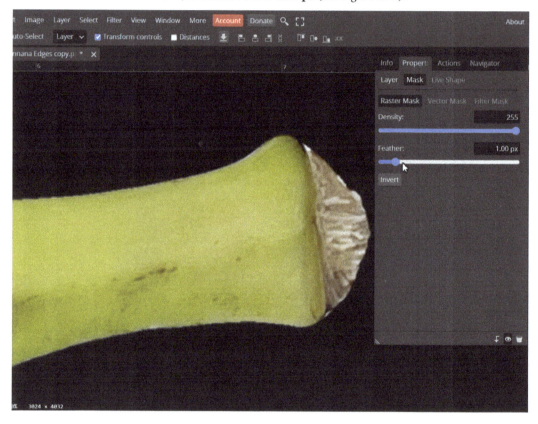

Figure 5.31 – Select Feather to smooth edges more

11. The **Feather** parameter allows us to soften the edges once a selection is made. This helps make the selection seamless and less obvious like a cut-out.

12. We have now created a soft edge around the banana, giving it a more natural look.

Let's dive deeper into softening and refining edges in the next section.

The Refine Edge tool

The **Refine Edge** tool is similar to the **Median** and **Feather** filters, but more powerful. It allows you to make an accurate selection of things that have complex edges such as hair, fur, grass, and so on. We will apply it to an image of a young lady for this exercise, but we need to remove her from the background first:

1. Grab the photo reference of the young lady titled **Tam Img**, located in the `Chapter 5 Resources` folder and drag it into the Photopea Main area.

2. Choose the **Quick Selection** tool and click on the woman's face to select her.

3. While the selection is active, switch to the Magic Wand tool and start adjusting the selection to fit along the edges more accurately. Hold the *Shift* key and click the Magic Wand to expand the selected area where required. Wherever you see an area where the selection is too far outside of the edge, hold down the *Alt* key while clicking the Magic Wand to expand the selection closer to the edges of the lady. See *Figure 5.32*:

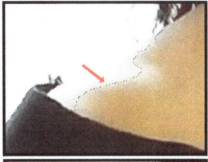

Figure 5.32 – Use Quick Selection to select and Magic Wand to adjust the selection

4. Once the selection looks more accurate, we can begin refining the selection with the **Refine Edge** tool.

> **Important note**
>
> The background will change to green here because I accidentally merged the image of the lady with a temporary green background. I thought this would be a good example of the mistakes that can happen with layers.

Let's look at the background change in *Figure 5.33*:

Figure 5.33 – Accidentally merged layers

5. Since the background color is neutral, it will be easy to select the lady with the **Quick Selection** tool again, and we can continue to refine the edges of the selection as follows.

6. With the young lady selected, press the **Refine Edge** button located just below the **Filter** menu. See *Figure 5.34*:

Figure 5.34 – Press the Refine Edge button while the young lady is selected

7. You will see a new window open showing the selection of the young lady greyed out on the left. See *Figure 5.35*:

Figure 5.35 – Refine Edge window

8. Just above the lady on the left is the brush for refining the hair and edges, followed by white, grey, and black buttons. See *Figure 5.36*:

Figure 5.36 – Refine Edges brush tools

9. Selecting the white button to paint the edges of the woman's hair will reveal the green background on the image to the right, as shown in *Figure 5.37*:

Figure 5.37 – Refine Edge white button

10. Selecting the grey button to paint the edges will improve the selection (which is what we want) for uncertain areas. See *Figure 5.38*:

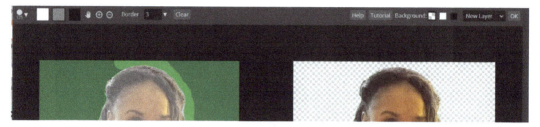

Figure 5.38 – Refine Edge grey button

11. Painting the edges with the black button will delete the woman. See *Figure 5.39*:

Figure 5.39 – Refine Edge black button

12. The window in full color to the right of the screen shows you the changes being made from edits on the left. You can add a temporary black background to see where parts of the background are showing through the hair and around the edges if desired. See *Figure 5.40*:

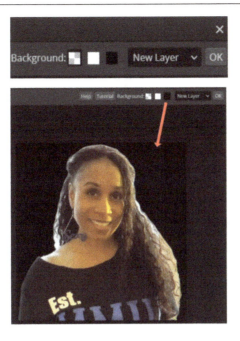

Figure 5.40 – Add a temporary background in the Refine Edge window to see areas to refine

13. Use the grey button to paint along the edges of the lady's hair, all along the skin, and the edges that need refining. (Don't paint too much into the image or object or it may alter the selection too much.)

14. Notice when you paint along an edge and let go, the right-hand window displays the changes. The green edges are removed and the edges of the hair are softened in the right-hand image. See *Figure 5.41*:

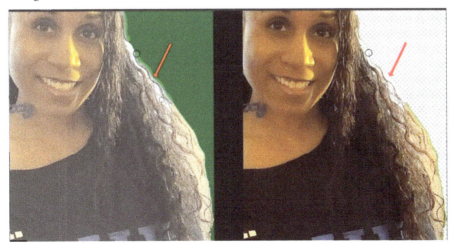

Figure 5.41 – Painting along the edges with the grey button

15. Once you've finished refining the edges, click the **OK** button on the top right of the **Refine Edge** window to close it. You will be back in the default work area.

16. **Refine Edge** has some limitations, so we will make further adjustments manually that will be similar but with a slightly different approach.

17. To do this, create a new **Raster Mask** on the layer of the image of the woman and make the foreground color bright green. Since the woman has dark hair and clothing, it will be easier to distinguish areas to touch up. See *Figure 5.42:*

Figure 5.42 – Create a new raster mask as a temporary background

> **Important note**
>
> This will act as a temporary background to help identify the white background showing through the woman's hair and around the edges of her shirt and skin that need to be cleaned up and softened manually with the brush tool.

18. Next, create a new Raster mask for the **Tam BK Flat** Layer. Click the small button with the *circle inside the square*, located at the bottom of the Layers panel, and a *white color box* will appear next to the **Tam Bk Flat** Layer.

This will allow us to manually remove the white bits between the tangled hair and soften the edges further. See *Figure 5.43:*

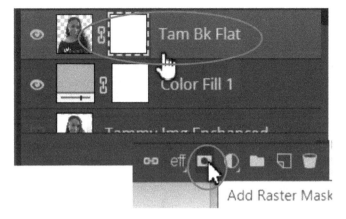

Figure 5.43 – New raster mask for refining the lady's hair

19. Select **Soft Brush** and set Brush **Size** to **4px**, **Opacity** to **18%**, and **Flow** to **100%**. Set **Foreground Color** to black, which will hide the white background as we paint over it, revealing what's hidden under the mask. See *Figure 5.44*:

Figure 5.44 – New adjustment layer for refining the lady's hair

20. Now we can begin painting in black over the edges of the hair to remove the white, hard, and jagged edges around the hair. The black shape you painted will show up as a shape on the mask. See *Figure 5.45*:

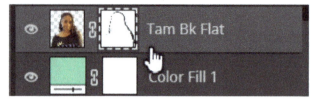

Figure 5.45 – Painting the mask to remove the white

You can see the white areas being removed nicely on the image close up in *Figure 5.46:*

Figure 5.46 – Painting in black to remove the white

21. Once we've finished painting in the black to remove the white areas, we can add a new layer above the woman and name it *Hair added layer*. We will use this layer to paint in a few strands of soft hair to make it look more natural. See *Figure 5.47:*

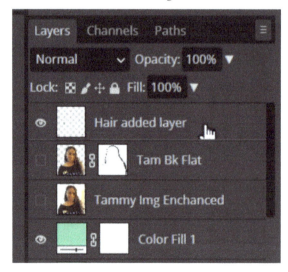

Figure 5.47 – Hair added layer

22. Next, use the **Eyedropper** tool to sample the hair color. See *Figure 5.48:*

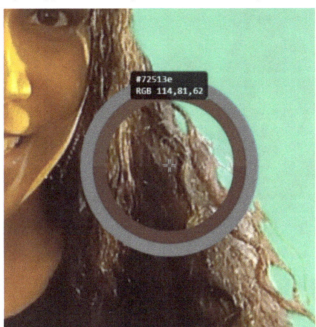

Figure 5.48 – Sampling the hair color with the Eyedropper tool

23. Select the **Soft Brush** and adjust **Opacity** to **18%** and **Flow** to **100%**, then trace in a few hair strands with the brush. See *Figure 5.49:*

Figure 5.49 – Select Soft Brush and adjust the Opacity and Flow values

24. I used the original photo layer title (**Tammy Img Enhanced**) as a reference when painting in the hair strands. By changing the opacity of the original photo to **39%** and placing it underneath the newly refined image, it shows parts of the hair that can be painted back in. See *Figure 5.50:*

Figure 5.50 – Original photo as a backdrop to help draw in hair strands

25. Let's zoom in on the hair to see the added strands of hair on the left and right images, compared to the center image showing the hair before the strands were painted in. See *Figure 5.51*:

Figure 5.51 – Drawing in hair – close-up view

26. I also painted in some of the brown skin on her arm by sampling a similar color with the **Eyedropper** tool. You can see the paint strokes I made for the hair and skin layer on the left in the following figure (shown against a green background for clarity). See *Figure 5.52*:

Figure 5.52 – Final image refined with hair strands drawn in on a separate layer

Now that we've covered this section on the refinement tools for edges and complex selections such as hair, we will learn a few other ways of removing backgrounds from an object or figure in the next section on channels.

Working with channels

Photopea and other popular image editing applications use channels that contain color information about an image and can preserve selections. All images comprise at least one channel, and often more. When Photopea opens an image, it automatically calculates and stores the color channels. Both channels and masks are greyscale images that you can edit like any other image.

The number of color channels created depends on an image's color mode. Each channel can be edited, deleted, or converted to a mask, which enables you to protect specific parts of an image. You can also opt to hide and show channels. Showing just one channel displays the image in grayscale.

Common color modes are RGB, CMYK, Bitmap, Grayscale, Duotone, and Indexed Color mode:

- The RGB color mode consists of three channels: Red, Green, and Blue, along with a composite channel for editing the main image

- The CMYK color mode consists of four channels: Cyan, Magenta, Yellow, and Black, but has a smaller gamut to simplify the four-color printing process

- The Bitmap, Grayscale, Duotone, and Indexed Color mode only create one channel

Now that we covered the common color modes associated with channels, let's see how we can access the **Channels** panel in the next section.

The Channels panel

The **Channels** panel is located on the right side of the workspace along with the **Layers** panel. If it's hidden, you can go to **Window** in the menu bar and select **Channels**: See *Figure 5.53*:

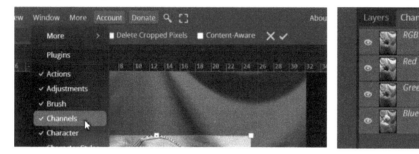

Figure 5.53 – Locate and select Channels

Now that we've covered the **Channels** panel, we can move on to creating a channel from a selection.

Create a channel from a selection

Understanding and learning to create channels from a selection will be very beneficial, adding flexibility to your workflow and saving you a lot of time in the long run. Let's find out how in this section:

1. Make a selection of the man in the blue hoodie, then click on the **Channels** tab to reveal the RGB channels. You can hide and reveal the channels to see how the colors behave.

2. At the bottom of the **Channels** panel, click **Save Selection as Channel** to create an Alpha channel (see *Figure 5.54*).

An Alpha channel can be saved permanently, so you are able to reload it, place it into another image or document, and also save it as an Alpha path for a 3D render.

Figure 5.54 – Create an Alpha channel

You can hide and reveal different channels to see how the colors behave. See *Figure 5.55*:

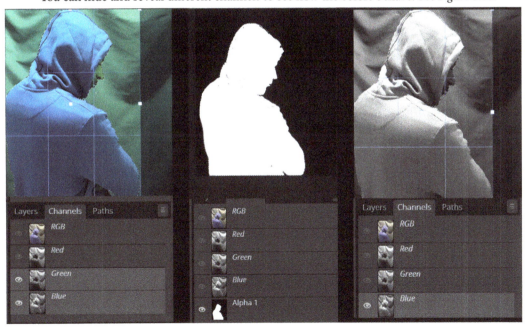

Figure 5.55 – Alpha channel and how colors behave

That concludes the end of this section on working with channels. Let's sum up all the things we covered in this chapter.

Summary

We covered a lot of information and techniques on various ways to make selections in Photopea. We learned how to identify and use a number of selection tools, namely the Lasso, Polygonal Lasso, Rectangle and Elliptical Select, and Magic Wand selection tools. We also learned how we can avoid merging objects together by mistake when making selections.

We saw how to edit our selections to make better selections using the Modify feature to adjust the selection, as well as the feathering feature. In addition to that, we explored the Replace, Unite, Subtract, and Intersect selection modes.

We learned how to soften the edges to give them a natural seamless finish, how to remove backgrounds, make complex selections of objects such as hair, and paint in strands of hair. Finally, we touched on the Pen tool and gained an understanding of how to work with channels.

That concludes this chapter. We are ready to move on to *Chapter 6, Color Theory and Application.*

Part 2: Digital Imaging, Design Techniques, and Other Software

This part takes a deeper dive into Photopea, exploring color theory and application. We will learn how to use swatches and the color picker and look at the Brush panel. We will also learn how to use brushes and create our own. We will touch up a headshot and create an image composite from multiple photos. We will also learn how to style paragraph text and choose the best font for a project. Lastly, we will learn how to customize Photopea's pre-designed templates and create an animation.

This part comprises the following chapters:

- *Chapter 6, Color Theory and Application*
- *Chapter 7, Using and Creating Brushes*
- *Chapter 8, Photo Retouching Techniques*
- *Chapter 9, Exploring Advanced Image Compositing Techniques*
- *Chapter 10, Text Fundamentals and Styling in Photopea*
- *Chapter 11, Pre-Designed Templates, Extra Effects, and Features Overview*

6
Color Theory and Application

In this chapter, we will begin by exploring and understanding the basics of color theory, design concepts, and principles that are involved in creating great compositions. Some of the things we'll examine are the color wheel, and examples of how color can express mood, and be applied with the Photopea application; such as using and creating swatches, choosing color profiles, using the color picker, and more.

By the end of this chapter, you'll have had a solid introduction to color theory that you can apply to your images and compositions with Photopea.

In this chapter, we will cover the following topics:

- Color theory basics
- Working with color panels
- Using swatches and the color picker
- Working with color spaces

Color theory basics

Not delving too deep, we will only touch on color theory lightly. I would encourage you to research color theory and the psychology of color if you want to expand your understanding and improve your work.

Color is one of the seven visual elements (line, shape, form, value, space, and texture) used for creating composition in art and design. Color has the most influence and effect on our emotions, mood, and physical state, compared to the other elements. There are also various aspects of color; for example, light, tone, pattern form, symbol, movement, harmony, contrast, and mood. We use color to create mood and atmosphere in art, graphic design, advertising, home decorating, fashion design, and so on.

How we see and process color

The eyes and brain work together to translate and create the images that we see in color. The retina is located at the back of the eye and is a part of the brain, and the central nervous system. The eye delivers visual information through the optic nerve, from the retina to the brain. In addition, the eyes have light receptors that transmit signals to the brain that are stimulated by color created by different wavelengths of light. Each wavelength determines how we perceive each color on the color spectrum (red, orange, green, etc.).

Some individuals are unable to distinguish as many color ranges as others due to color blindness (a common color impairment). Let's break down some of the emotions that colors can trigger.

The emotions of color:

- Blue: Creates a sense of calmness, peace, harmony, trust, and reliability.
- Green: Can create a sense of nature, calmness, hope, and healing.
- Yellow: Can encourage a sense of excitement, warmth, abundance, and caution.
- Purple: Can create a sense of calmness, royalty, wisdom, nobility, and luxury.
- Red: Can increase enthusiasm, respiration, and blood pressure, and encourages action.
- Pink: Similar to red, increases blood pressure, respiration, and enthusiasm. It can also be used to reduce erratic behavior.
- Orange: Can stimulate activity, energize, and give a sense of warmth and desire.
- Gray: Can produce a sense of balance, calmness, strength, maturity, and security.
- Black: Gives a sense of mystery, power, and class; stimulates strong emotion.
- White: A sense of cleanliness, purification, and hope – encourages mental clarity.
- Brown: Creates a sense of stability and steadiness – natural and genuine.

Let's take a look at the traditional color wheel and a few common color schemes in the next section.

The color wheel

Having a physical or digital copy of a color wheel will be a great guide for choosing color schemes and a reminder of how colors relate to one another. I like to use a color wheel by *The Color Wheel Company*. It has two sides, with color theory and color scheme notes.

You should be able to find it at your local arts and craft stores and can find it online at www. colorwheelco.com/buy-now/product/creative-color-wheel/. You'll find a variety of color wheels for specific uses and purposes. You can also google and find plenty of color wheels to download on your computer or smart device.

I would also recommend that you create your own color wheel and color schemes in a sketchbook, or create them in image editing software such as Photopea, to help further your learning and understanding of color psychology and mixing colors.

This is a color wheel I created to give you an idea. See *Figure 6.1*:

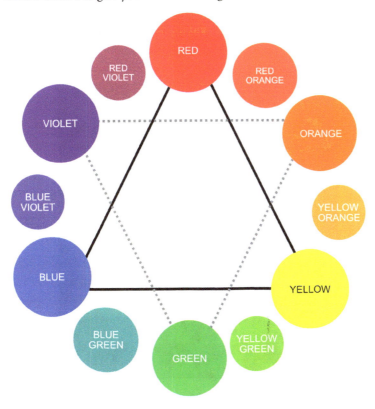

Figure 6.1 – Color wheel example

Now that we've covered the color wheel, let's move on to the next section regarding color.

Color schemes

Understanding the basics of color, and learning how to mix colors for color schemes can enhance your work, and help communicate and express the mood and voice of your designs, and paintings. Let's look at the different color schemes we can apply to our creations:

- **Primary colors**: Red, Yellow, and Blue – they are the three essential colors used for mixing all hues (plus black and white).

- **Secondary colors**: Created by mixing two primary colors – for example, Red + Yellow = Orange, Blue + Yellow = Green, Red + Blue = Purple.

- **Complementary colors**:

  - These are colors that contrast with each other, set up diagonally opposite each other on the color wheel.

  - They create the strongest contrast when applied to images or compositions.

  - Create one color to be the dominant and the second one to be the accent color for good contrast and balance.

  See *Figure 6.2*:

Figure 6.2 – Complementary color scheme

- **Tertiary colors**:

  - These are created when primary and secondary colors are mixed adjacent to each other on the color wheel, for example, Red Purple + Purple Blue + Blue Green + Green Yellow, and so on.

- **Analogous colors**:

  - These are colors that sit next to each other on the color wheel, for example, Red, Pink, and Purple are next to each other. See *Figure 6.3*:

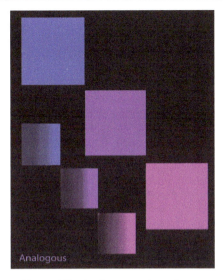

Figure 6.3 – Analogous color scheme

- **Monochromatic colors**:

  - These are created from one color that can create a wide range of colors, from a variety of different shades, saturation, and intensity within the same base color. For example, mixing a small amount of dark green with a brighter green. The more dark green you add to the brightest green, the darker and less saturated it will become. See *Figure 6.4*:

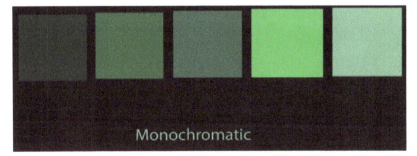

Figure 6.4 – Monochromatic color scheme

Now that we've covered some basic color schemes, let's dive into the next section, *Color theory terms*.

Color theory terms

Learning and gaining an understanding of color theory terms helps us communicate verbally, as well as being intentional when applying them to our images:

- **Hue**: This is one of many colors of the spectrum. The visible colors of the spectrum are red, yellow, orange, green, blue, indigo, and violet.

- **Saturation**: This defines how pure and intense a color is.

- **Value**: This refers to how light or dark a color is.

That wraps up this section. I recommend studying color theory further than what we've covered here.

Working with color panels

Both the **Color** and **Swatches** panels are located in the sidebar (2) (on the left side, within the two vertical columns. You may have to expand the columns by clicking the small < or > icon (1) if the **Color** panel is hidden. The **Layers**, **Channels**, and **Adjustments** tabs are also located in the **Sidebar**. You can also find the **Color** and **Swatches** panel under the **Window** menu located just above the workspace. See *Figure 6.5*:

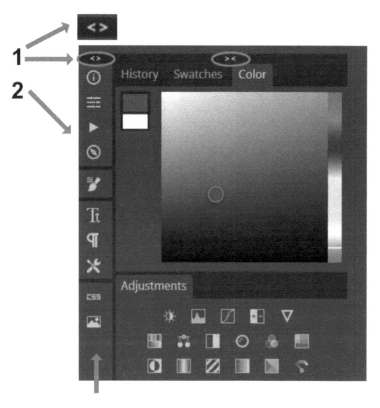

Figure 6.5 – Locating the Color panel

Once the **Color** panel is open, you will see the active (foreground) color displayed in gradations of white to dark. If you slide it to the left, you can make the color (an orange tint) less saturated by adding white. The further you drag it toward white, the less orange pigment will appear. If you drag the color picker toward black, the darker the orange will become (shades of orange), and it will eventually become a solid black color if you continue to drag it toward black. See *Figure 6.6*:

Figure 6.6 – Dragging the Color panel range from white to black

On the right of the **Color** panel, you will see a vertical full spectrum of colors. You can change the color in the palette window by clicking and dragging the slider up or down. See *Figure 6.7*:

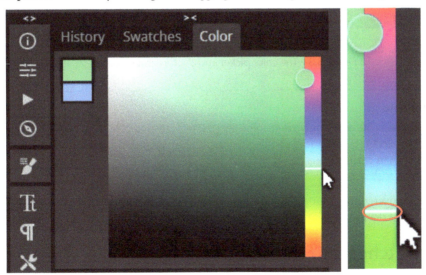

Figure 6.7 – Changing the color in the Color panel

To the left of the **Color Slider**, you will notice two squares of color. These are the foreground and background colors you see on the lower-left **Toolbar**. If you double-click on the green color, a second window (the color picker) pops up.

As you make adjustments in the **Color** panel, it will also change the foreground color, located in the lower-left toolbox. See *Figure 6.8*:

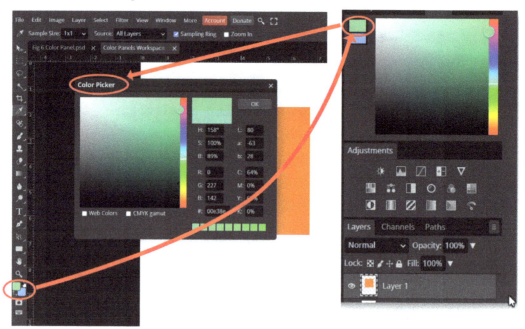

Figure 6.8 – The color picker and the foreground color

Important note

The foreground is the active or current color, and the background color is the layer background color. If you select the Eraser tool and begin erasing the background color, it will paint in the color, rather than erasing it.

Now that we've covered the color picker and foreground color, let's break down using swatches and the color picker in the upcoming section.

Using swatches and the color picker

As I stated earlier, the **Swatches** panel is also located in the **Sidebar**, and also under the **Window** menu. It displays the default colors as a quick start, and will also store the colors you work with regularly. See *Figure 6.9*:

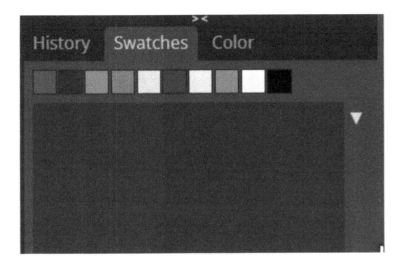

Figure 6.9 – Changing the color in the Swatches panel

You can remove or add colors from the **Swatches** panel, create and save a custom set of swatches as a library, create a custom color swatch from an image, or load a library swatch such as **Pantone Coated** or **Uncoated Color** libraries from other programs such as Photoshop.

Applying colors to shapes and images with swatches

Keeping your colors consistent in your images and compositions can save you time, trouble, sometimes money, and headaches from clients who use specific color schemes and swatches.

Learning how to sample, select, and save color schemes can save you the trouble and frustration of trying to mix the same color every time you open or create a new design or image. We will do that and more in the upcoming sections. Let's get started with *Managing Swatches*.

Managing swatches

Let's first look at adding color(s) to the **Swatches** panel:

1. To add a color to the **Swatches** panel, simply click on a color in the palette.

2. Next, double-click on the foreground color on the lower-left toolbar.

3. Create a new color with the slider and the new color will appear in the **Swatches** panel. See *Figure 6.10*:

Figure 6.10 – Create a new color in the Swatches panel

4. Another way to add color is by selecting a color from an image using the **Eyedropper** tool: Select **Eyedropper**, place it over an area of the image and click it with the mouse. A new color will appear in the **Swatches** panel. See *Figure 6.11*:

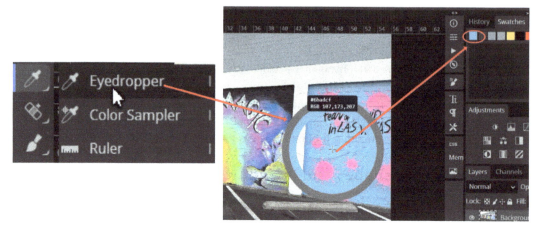

Figure 6.11 – Adding a color using the Eyedropper tool

Now that we've covered adding colors to shapes and images to swatches, let's dive into the next section, *Deleting color (s) from the Swatches panel*.

Deleting color(s) from the Swatches panel

To get started with deleting colors, *select* the color you wish to delete.

Next, click the small upside-down triangle in the right corner of the **Swatches** panel and click **Delete**. See *Figure 6.12*:

Figure 6.12– Delete a color from the Swatches panel

Now that we know how to delete a color From the **Swatches** panel, let's look at the color picker and **Eyedropper** tool.

Color picker and Eyedropper tool

When used individually and simultaneously, the color picker and **Eyedropper** tools can make it easy to sample and save colors, and also to create new ones that are in harmony with the initial colors that were included in an image. Let's briefly explore what these tools are, and how to apply them to your project:

- **Color picker tool**: This tool allows you to adjust, edit, or alter a foreground color.

- **Eyedropper tool**: This allows you to sample a color from an image and use it for your own projects.

- **Custom swatches from an image**: One of the benefits of creating custom palettes is keeping your colors consistent without guessing or manually recreating the colors to match your project consistently while you're working. Another benefit is being able to share the palette with another person working on the project, or sending it to a printer for consistency.

In this next example, we will go through the steps to make a custom swatch palette:

1. To do this, open an image or my image titled **Vegas Colors**.

2. Use the **Eyedropper** tool to select the light blue color of the sky (it will appear in the **Swatches** tab, as seen in *Figure 6.13*):

Figure 6.13 – Selecting the color of the sky with the Eyedropper tool

3. Next, select the water and other areas with the **Eyedropper** tool, until you have roughly 4-8 colors sampled from the image. See *Figure 6.14*::

Figure 6.14 – Continue selecting color samples with the Eyedropper tool

You will see all of the colors sampled with the **Eyedropper** tool in the **Swatches** panel. See *Figure 6.15*:

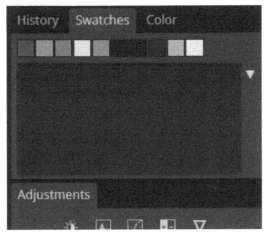

Figure 6.15 – All of the main colors sampled with the Eyedropper tool

4. Now that we've sampled the colors in the **Swatches** palette, we can begin inserting the colors one by one into the panel tray.

5. Note: swatches will only be added when we see the **Swatches added.** popup window – see *Figure 6.16*:

Figure 6.16 – Swatches added alert

6. Select the first color, then click the white down arrow on the right of the **Swatches** box and select **Define New**. The color will appear in the rectangular space just below the colors sampled, and a small **Default** folder will automatically be created for the new **Swatches** color library. Continue this until all the swatches are added. See *Figure 6.17*:

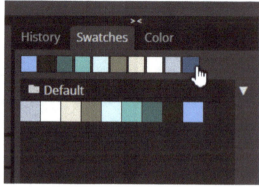

Figure 6.17 – Using Define New to add each color to a new Swatches collection

Now that we've reviewed how to define new colors and add each color to a new **Swatches** collection, let's jump into the next section, *Loading (importing) swatches into the panel*.

Loading (importing) swatches into the panel

There may be an instance where you've closed Photopea and the swatches you saved aren't visible in the **Swatches** panel. You can retrieve them by clicking on the white upside-down triangle on the right side of the Color panel and selecting **Open.ACO** (**ACO** stands for **Adobe Photoshop Color** file).

You need to browse through your computer's folders, look for the **Swatches** panel file icon, and select it. Photopea usually saves this file and any other files you've altered or exported as documents in the computer's download folder. See *Figure 6.18*:

Figure 6.18 – Load saved swatches back into the panel if not visible

> **Important note**
> I would suggest renaming the swatch so it's easier to distinguish between other swatches you create over time.

See *Figure 6.19*:

Figure 6.19 – Rename saved swatches

Now that we've covered renaming saved swatches, let's learn how to back up swatches with layers and shapes.

Backing up swatches with layers and shapes

I am also going to make the same palette from the Las Vegas photo with the **Rectangle Shape** tool to create each color on a Layer. Afterward, we can group the rectangle colors and flatten them into one Layer.

This will act as a backup of the swatch in case the program has a glitch and doesn't save the swatch palette in the panel:

1. To begin, select the **Rectangle** shape, create a shape, and fill it with a color from our custom pallet (see *Figure 6.20*).

2. The rectangle will be displayed on a single layer titled **Shape**. Each shape will be named **Shape 1**, **Shape 2**, and so on. You can rename them using the name of the color – for example, navy shape or green shape, but it is not necessary if you plan to group and flatten the palette as one layer.

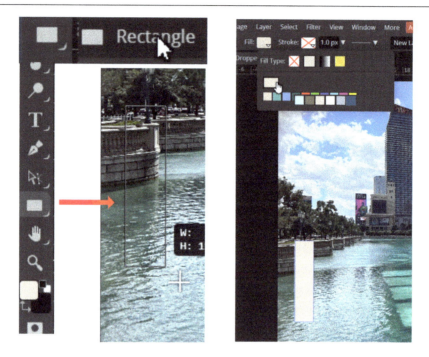

Figure 6.20 – Create a rectangle and fill it with the sampled color

3. Next, duplicate the layer rectangle, select the next color, and drag the second rectangle to the right of the first color. Repeat this process until we have created the palette on all the Layers in the document. See *Figure 6.21*:

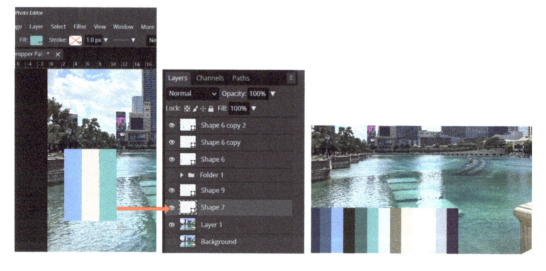

Figure 6.21 – Continue duplicating the rectangle and fill it with the next sample color

4. We can export this palette as an image, JPG, and so on. See *Figure 6.22*:

Figure 6.22 – Finished color swatch created with the rectangle tool

That covers the basics of backing up swatches with layers and shapes. Let's move on to the next section.

Working with color spaces

Color spaces can be a bit complex at first, when you look at different color models, math, theory, technicalities, and so on, but will get easier to grasp as you continue to work in photo editing programs such as Photopea and Photoshop.

Color spaces in Photopea

To create a document with a particular color space in Photopea, we will execute the following steps:

1. Go to **New Document** and click on the small down arrow to the right of the **Background** color. See *Figure 6.23*:

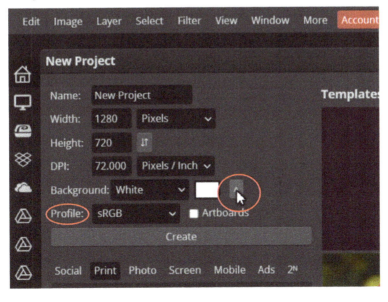

Figure 6.23 – Create a new document and select a color space

2. A sub-tab will open called **Profile**: Click on **sRGB** to access the four color spaces to choose from according to which of those best suits the needs of your project: **sRGB**, **Adobe RGB**, **ProPhoto RGB**, and **Display P3**. See *Figure 6.24*:

Figure 6.24 – Click on sRGB to select a color space

> **Important note**
>
> At the moment, Photopea can only set up a color space when creating a new document. Hopefully, they will update it to change the color spaces of existing files sooner rather than later.

I will still break down some of the basics so that you're aware and able to apply them to other image editing programs that have the option to change the color space.

Color spaces can be better understood when we understand what color is; which we covered a little bit in the *Color theory basics* section (you can go back and review it). We know that our eyes can pick up more colors than standard smart devices, and the number of colors varies per device.

A color space is the range of colors available on your capturing devices (digital cameras, scanners, and drones (some of these are also available on smartphones) and display devices such as printers, computers, and television monitors.

Although the colors may display slightly differently on display devices, color spaces provide a consistent system to produce and preserve the same colors across a variety of devices.

Some common color spaces are **sRGB**, **Adobe RGB**, **ProPhoto RGB**, and **Display P3**. Let's learn more about them next:

- **sRGB**: This is the smallest color space that produces the least number of colors. Web browsers and devices tend to use this profile more, considering some smart devices display more or fewer colors than others. This ensures we all view the same colors consistently.

- **Adobe RGB (1998)**: This produces a mid-range of colors. It displays more colors than sRGB, but will not be able to display all of the color information from a RAW file.

- **ProRGB**: The higher the color space range, the more saturated color ranges will be included.

  Since sRGB does not display Adobe RGB and ProRGB, you will have to convert your color space to sRGB for it to display on the web correctly. When you export the converted color space, do not save the original working file as sRGB or permanent color loss will occur.

- **Display P3**: This is a color space developed by Apple Inc. It was developed from the DCI-P3 color gamut found in digital cinema projectors; but uses a slightly different version of white identified as D65. It offers more greens and reds than the standard sRGB color gamut. Though it's larger and provides more color space than sRGB, it's still not large enough to cover color for professional environments.

> **Important note**
>
> When converting your images to different color spaces in programs such as Photoshop, and Affinity Photo, be mindful that permanent color loss can occur, depending on which mode you choose. Be sure to save backups of your original image until you are sure of the results.
>
> Photopea has some limitations in converting images to Color Spaces. For example, you would have to create a new document and click Profile: You can choose from sRGB, Adobe RGB ProPhoto RGB, and Display P3.
>
> Once the new document is set up, you can then drag your image into that document's color profile.

Standard color spaces

Since we know our eyes can view more colors than standard devices and internet browsers, let's take a brief look at how the color spaces we perceive are calculated and interpreted with two standard color spaces: **CIELAB** and **CIEXYZ**.

> **Important note**
>
> RGB stays consistent with the three primary colors in the form of light. Because it uses light to emit colors, you could end up with too many of the same color under different number sequences. For example, orange mixed from red and yellow is one version, but you can also create a similar orange by mixing different shades of orange with other yellows, and one other color. This would make it overly complex, and too large to contain all of the color information.

That's why **CIE 1931** was developed; it is a color-matching system based on math that gives calculations based on how colors are viewed by the average person for reproducing colors for printing purposes. It can be used to manipulate RGB into absolute colors, and it's able to put similar colors under the same number sequence, reducing the chances of the same repeated colors under different number sequences.

CIELAB color space, also called La^*b^*, is designed to give you the closest resemblance to a color viewed by the human eye.

The L^* component closely matches human perception of light and can calculate subtle color differences. The a^* and b^* represent the four unique colors of human vision: red, green, blue, and yellow.

In *CIELAB*, a numerical change is translated from a color change. These results are not based on any specific computer monitor or printer. *CIELAB* (also CIEXYZ space) is an independent standard observer models that give you an average of results from color-matching experiments held in laboratories.

The *CIE 1938* and *CIE 1931 XYZ* color spaces were created by the International Commission on Illumination. The CIE XYZ color space involves all color sensations visible to someone with average eye vision. It serves as a standard reference for color spaces.

Here is the meaning of *XYZ*:

XYZ represents the *three cones inside the human eye*. Z = *the S Cone*, Y = *a* combination of the L and M responses, and the X value is a *mix of all three*.

Y stands for Luminance, Z is quasi-equal to blue (of CIE RGB), and X is a mix of the three *CIE RGB* curves.

Color modes (also called color models) serve a different purpose. They factor out how colors are defined in numerical values, whereas color space represents the gamut of colors displayed.

In addition to color modes (models), RGB, CMYK, and Grayscale are the most common color modes. Both RGB and CMYK color modes work differently to produce the colors we see in images and will dictate how we approach and set up different projects for the web, display, printing, and so on.

> **Important note**
>
> Color modes, color spaces, and color profiles are not the same thing; each serves a different purpose, yet works together interdependently to manage color.

Color profiles: A color profile is embedded with data that can depict a particular color space, such as RGB, or hardware devices such as a scanner, camera, and so on. In most instances, color profiles are created as an ICC profile. They are small files that have an a `.icc` or `.icm` file extension. A color profile can be implanted into an image to define or clarify the gamut range of data. This provides a more comprehensive system that allows users to see the same colors across different devices without losing or modifying colors.

That sums up the essentials of color profiles.

Summary

In this chapter, we covered a lot regarding color theory and how to apply it to the powerful tools and workflows within Photopea. In doing so, we gained an understanding of how we perceive colors with our eyes and brain and the psychological effects color can have on our minds, moods, and emotions. We also learned how to navigate and locate different tabs for color swatches and panels, and how to customize swatches with tools such as the eyedropper.

In the next chapter, we will learn about the tool brush, explore some different types of brushes, and learn how to create our own for touching up photos, drawing, and painting.

7

Using and Creating Brushes

In this chapter, you will gain a solid understanding of what the brushes in Photopea are capable of achieving in your projects. In addition to that, we will look at the **Brush** panel, presets, patterns, and how to import and create our own brushes. By the end of this chapter, you will know how to access brushes, create custom brushes, and manage brushes to tackle the upcoming projects in *Chapter 8, Photo Retouching Techniques, Chapter 9, Exploring Advanced Image Compositing Techniques*, and other areas of this book.

For this chapter, we will cover the following topics:

- Exploring and accessing the brushes
- Using the default brushes
- Editing default brushes with custom Dynamics
- Creating custom brushes from images

Exploring and accessing the brushes

As we continue to navigate Photopea, we discover this open-source photo editing software is packed with powerful tools for those working at a beginner level to a professional level. If you plan on using photo editing long-term, then you need to grasp the essentials of working with the Brush Tool and the basic tools that aid them.

Both the Brush Tool and Pencil Tool are your digital Tools, which give you unlimited possibilities and flexibility. They are used much like traditional paintbrushes and pencils for drawing, painting, and making marks on paper, canvas, and other surfaces. Keep in mind that the Brush tool has a variety of functions and settings that react to layers (or the canvas), each with a different or unique result.

We will start with the basics and gradually experiment with some of the functions and settings to keep it simple and interesting. I encourage you to continue exploring the Brush tool beyond what is covered in this chapter.

Now, let's explore the **Brush Tool** in Photopea and see how it can enhance your photo editing experience.

Accessing the Brush tool

To get started, let's select **New Document** and set the screen size to 1920 x 1080 px, or any size you prefer.

Next, the **Brush Tool** is located in the Toolbar, on the left side of Photopea's document window (see *Figure 7.1*). You can also press *B* on the keyboard to quickly access the **Brush Tool** and easily get started using the preset brushes that come with Photopea.

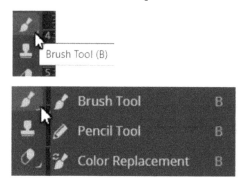

Figure 7.1 – Accessing Brush Tool in Toolbar

You will notice the **Brush Tool** parameters and settings (just below the **Top** menu) for the **Size**, **Blend Mode**, **Opacity**, **Flow**, and **Smooth** options, which you can change anytime you feel the need to. See *Figure 7.1*:

Figure 7.2 – The Brush Tool parameters

The supporting tools, such as the **Eraser Tool**, **Blur Tool**, and **Smudge Tool**, are also located in the Toolbar:

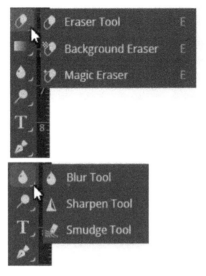

Figure 7.3 – Accessing Eraser Tool, Smudge Tool, and Blur Tool

This covers how you can access the **Eraser Tool**, **Smudge Tool**, and **Blur Tool**. Now, we can move on to the next section, *Selecting the Brush tool*.

Selecting the Brush tool

When you select the brush, click the **Brush Size** arrow button to access the current brush's **Size** and **Hardness** options, as well as the brush library, to easily change the current brush tip and type to a different **Default Brush**. Also, if you drag the slider down, you can see other brush tools and brush packs that you may have downloaded and installed from other third-party sources into Photopea. See *Figure 7.4*:

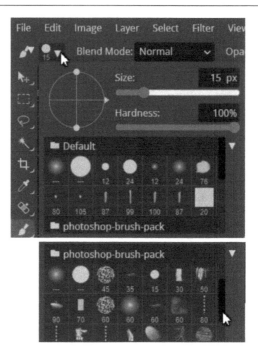

Figure 7.4 – Brush Tool's Size, Hardness, and Brush Library

Just below it, you'll see the **Default** Brush library. On the right of the **Default** Brushes tab is another tab for managing brushes, including **Define New** (create new brushes), and **Load.ABR** (import brushes), **Export as.ABR** (export brushes), **Name Change** (rename brushes), and **Delete** options. See *Figure 7.5*:

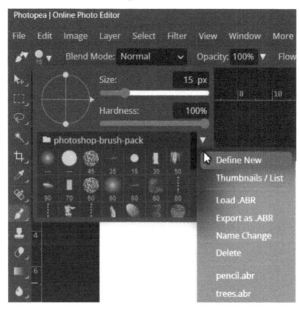

Figure 7.5 – Sub-tab for managing brushes

Now that we've covered how to access and select the brushes, let's move on to the next section.

Changing the brush color

It's easy to change the brush color by clicking on the foreground or background color located at the bottom of the Toolbar.

Decide which color will be the foreground color and which will be the background color.

Just below the Foreground Color option, click on the small *90*-degree arrow to switch the color from the foreground to the background. See *Figure 7.6*:

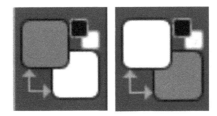

Figure 7.6 – Changing the brush color with the foreground and background button

You can change the colors for the Brush by double-clicking on the foreground color to access the **Color Picker** settings, and dragging it, or, you can click in the color range for more color options. See *Figure 7.7*:

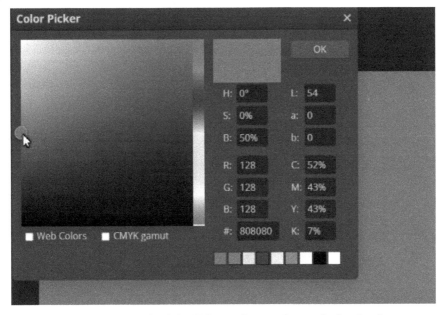

Figure 7.7 – Using the Color Picker settings to change the brush color

Now that we've covered how you can change the brush color, we are ready to dive into the next section, which is about the **Brush** panel.

The Brush panel

You can access more settings and presets for brushes in the **Brush** panel.

Go to **Window** and select **Brush**. You will see advanced presets and controls for the brush tip (see *Figure 7.8*), which include **Tip Shape**, **Tip Dynamics**, **Scatter**, **Color Dynamics**, **Transfer**, **Size**, **Angle Spacing**, and **Roundness**. We will go into more detail with the **Brush** panel in the *Editing default brushes with custom Dynamics*, *Managing brushes*, and *Creating custom brushes from images* sections, and ongoing throughout this chapter.

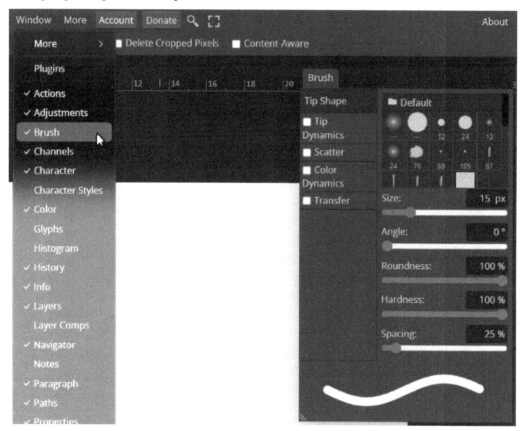

Figure 7.8 – The Brush panel

Now that we've gained a basic understanding of how to access **Brush Tool** and change brush colors, as well as become familiar with the presets, we are ready to start using the brush in the next sections.

Using the default brushes

There are a variety of brush tips and textures: hard, medium, and soft brushes, ink pen, airbrush, charcoal, paintbrush, and more. We will cover a variety of default brushes to experiment with mark-making and drawing and get a feel for how each brush interacts with the canvas (layer). See *Figure 7.9*.

This is a good chapter for beginners, as we will use the brushes for a variety of tasks and projects throughout the book. As we cover topics such as editing the default brushes, managing brushes, and editing the default brushes with custom **Dynamics** throughout this chapter, the brushes, and how to use them, will make more sense. I encourage you to experiment on your own, but learn how to use the default brushes first; specifically, the basic **Hard Round** brush and the **Soft Round** brushes, and gradually expand from there.

Figure 7.9 – Experimenting with brushes

> **Important note**
> You can check out https://www.xp-pen.com, which has some affordable tablets that are great alternatives to the more expensive Wacom Intuos tablets.

Experimenting with the default brushes

To get started using the default brushes, select **Brush Tool** from the **Toolbar** menu.

Select the **Hard Round** brush, set the **Hardness** value to **100%** and the **Brush Size** value to **150 px**, and draw on a small area of the canvas. See *Figure 7.10*:

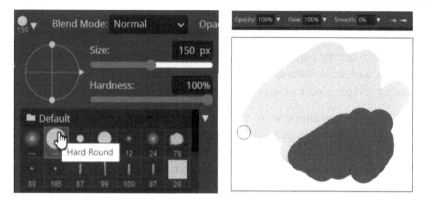

Figure 7.10 – Using the Hard Round brush

Next, select the **Soft Round** brush, set the **Hardness** value to **100 px** and the **Brush Size** value to **400 px**, and draw on a small area of the canvas. See *Figure 7.11*:

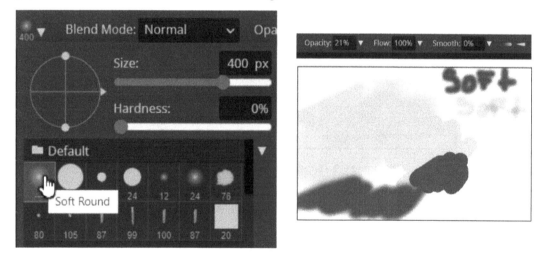

Figure 7.11 – Using the Soft Round brush

Next, create a new layer by selecting **New Layer** and experiment with some of the other brushes, such as the **Charcoal**, **Water Color**, **Soft Square**, and **Rough** brushes. This will get you comfortable with accessing and using the brushes and may generate ideas for future projects. See *Figure 7.12*:

Figure 7.12 – Trying out different paintbrush tips

Now that we've experimented with mark-making using various Photopea default brushes, let's jump into the next section, *The brush settings*.

The brush settings

Learning how to adjust the brush settings for different tasks and projects will be a huge benefit for solving creative solutions, as well as understanding how to customize and create your own brushes down the line.

Experimenting with the default brushes

You can adjust the brush size using the square bracket keys (*[* and *]*) on the keyboard.

Selecting the left bracket key will decrease the brush size, and the right bracket key will increase the brush size.

You can also change the brush size by clicking on the brush size icon near the top toolbar and dragging the brush size slider left to decrease the size or right to increase the brush size. See *Figure 7.13*:

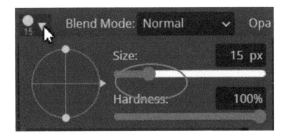

Figure 7.13 – Changing the brush size

Now that we've covered how to change the brush size, let's move on to the *Opacity* section.

Opacity

The brush's **Opacity** setting allows you to control how saturated the color displays on the canvas. The higher the **Opacity** setting, the more vibrant and concentrated the colors will be. The lower the **Opacity** setting, the longer it takes to build color up. It's a slow, subtle process that may require you to make more brush strokes to get it to its darkest or saturated color. It's a great way to blend in different colors and values (subtly going from dark to light). See *Figure 7.14*.

The following figure is an example of the **Hard Round** brush changing the **Opacity** setting:

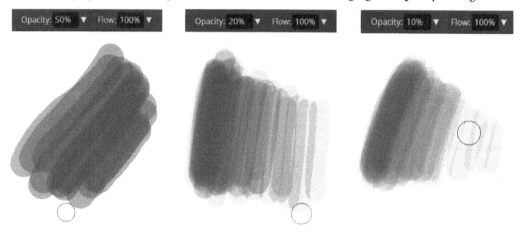

Figure 7.14 – Changing the brush's Opacity setting

That covers how you can change the brush's **Opacity** setting. Now, let's move on to the *Brush Flow*, *Brush Hardness*, and *Brush Mode* sections.

Brush Flow

The **Brush Flow** setting determines how fast the effects are applied from each brush stroke. The higher settings will be more dominant and apparent, while the lower settings will build up subtly; similar to building up strokes with the **Opacity** settings.

Brush hardness

Brushes with harder tips, such as **Hard Round** and **Hard Mechanical** in Photopea's default brushes, will lay down sharper edges. You can use these brushes for a variety of things, including drawing out custom fonts and lettering and line drawings for coloring books.

Brushes in programs such as Photoshop may have ink pens, mechanical pencils, and calligraphy pens that have hard, sharp-edged brushes.

These types of brushes are great for line drawings, coloring books, and adding variations to brush strokes with medium and soft brush tips.

Blend Mode

The brush's **Blend Mode** setting has a variety of brush stroke effects that you can apply to an image. Each effect may interact with different results.

In the example with the banana, I show how each **Blend Mode** setting can give you a different result from a brush stroke painted over the banana. See *Figure 7.15*:

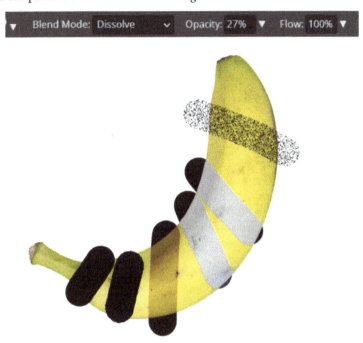

Figure 7.15 – Changing the brush Blend Mode setting

In this second example, I show how different **Blend Mode** settings interact with the canvas. You get more variations when changing the **Opacity** and **Flow** settings with each **Blend Mode** setting. See *Figure 7.16*:

Figure 7.16 – Changing the brush's Blend Mode, Opacity, and Flow settings

Now that we've covered the basics of how to change the brush's **Blend Mode**, **Opacity**, and **Flow** settings, let's begin the *Editing default brushes with the custom Dynamics* section.

Editing default brushes with custom Dynamics

The brush dynamics are the make and dimensions of a specific brush, which involve the **Brush Angle**, **Roundness**, and **Spacing** settings (the spacing or gap between a line or pattern of the brush's stroke).

Let's take a look at some key terms and settings to further our understanding of brush dynamics:

- **Tip Shape**: The brush shape is self-explanatory. You can determine the shape of the brush tip, such as round-shaped, oval, square, star, or any custom-shaped brush you create.

- **Spacing**: Setting the spacing at **25%** or less will make a solid line (continuous) while spacing over **25%** will have open space or gaps between the strokes.

- **Angle**: It enables you to change the angle of the brush.

- **Tip Dynamics**: This enables you to add random variations of the brush size, angle, and roundness with each stroke.

  - Select **Tip Dynamics** in the **Brush** panel to open the **Jitter Controls** sub-tab.

  - It gives you more control over the colors of your brush strokes with values you manually set.

- **Transfer**: The **Transfer** option creates a random effect for the brush's opacity and flow range (this is also the case for Flow and Jitter).

- **Opacity Jitter**: The **Opacity** parameter can change the brush strokes opacity, sporadically from higher to lower.

- **Scatter**: This allows you to arrange shapes spaciously near the area of the initial stroke created.

  In the following example, the **Angle** setting of the first paint stroke was **0 degrees** and **Spacing** was set at **1%**.

- The **Spacing** setting of the second paint stroke was set at **115%**.

- The **Angle** setting of the third paint stroke is **106** degrees, and the **Spacing** setting is **176%**, and the added **Tip Dynamics** and **Scatter** settings made the circles spread out as I dragged the brush horizontally across the canvas. See *Figure 7.17*:

Figure 7.17 – Tip Dynamics – Scatter and Spacing

- **Color Dynamics**: This allows you to edit the stroke color of each shape randomly.

  Rather than show examples of all the brush **Dynamics**, you can look at this example of me experimenting with some of the brush stoke variations when I select different **Tip Shape** settings and manually adjust settings for **Jitter**, **Scatter**, **Transfer**, **Color Dynamics**, and so on (see *Figure 7.18*).

We will go over brush dynamics more in the *Creating custom brush editing presets* section.

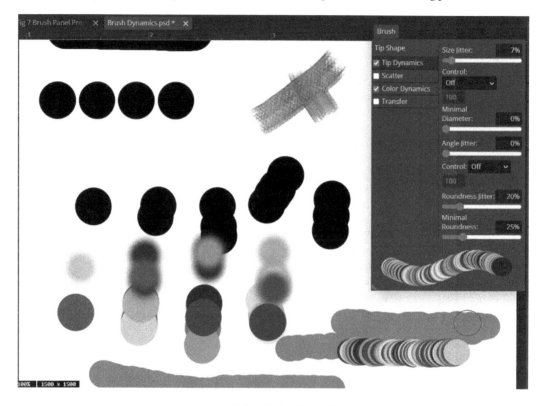

Figure 7.18 – Color Dynamics

Have fun experimenting; see what you come up with.

Creating custom brush editing presets

Create a new document with **1500 px x 1500 px** and use black as the foreground color.

Select the default **Hard Round Brush**. I created a brush stroke to compare the shape and presets after making the adjustments as seen in *Figure 7.19*:

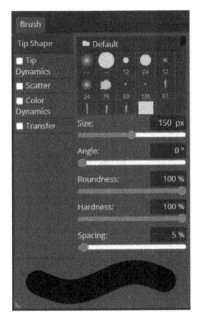
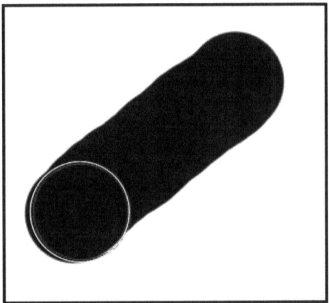

Figure 7.19 – A brush stroke with a Hard Round brush

Open the **Brush** panel and make edits and adjustments to the **Hardness**, **Shape** (**Roundness**), and the preferred **Tip Dynamics** settings.

For my example, I changed the **Hardness** setting from **100 %** to **50 %**, and the **Angle** setting from **0 % to 35 %**. See *Figure 7.20*:

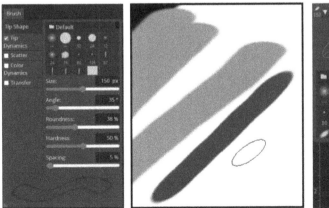

Figure 7.20 – Defining a new edited preset brush

Next, while the image is selected, go to the **Edit** menu, choose **Define New**, and select **Brush**.

The brush is created and ready to use. You can find the new brush in the **Brush** panel. See *Figure 7.21*:

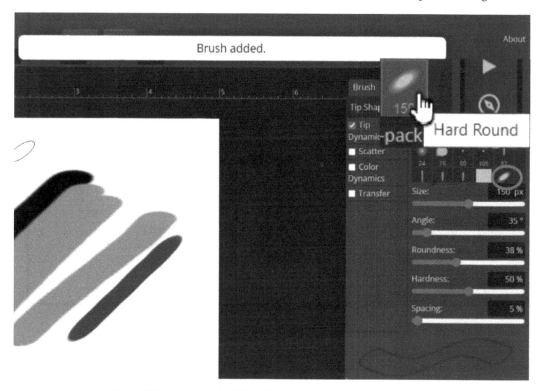

Figure 7.21 – The newly defined brush located in the Brush panel

Now that we've covered how to make a new brush by editing a default Brush, we are ready to cover the next section on creating custom Brushes from images.

Creating custom brushes from images

Creating custom brushes can be a lot of fun, as well as rewarding. You can use objects and other subjects that you may have photographed or found on stock image websites, and combine different objects from images to make custom brushes, giving you limitless possibilities. Let's create our very own brush from an image step by step:

1. Find some photos you would like to sample. For example: clouds, leaves, grass, or rocks.

2. Create a new document with a size of **1500 px x 1500 px**.

3. You can drag your image onto the newly created document and *resize* the image to fit slightly smaller within the **1500 px x 1500 px** template.

4. Next, while the image is selected (with the **Selection** Tool), go to the **Edit** menu, choose **Define New**, and select **Brush**.

The image is created as a **Brush** instantly and activated for use. You can find the new brush in the **Brush** panel. See *Figure 7.22*:

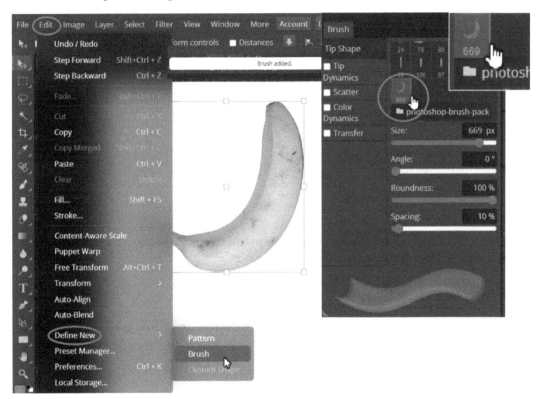

Figure 7.22 – Create custom brushes from images

5. You can begin painting with the **Brush** and adjust its parameters as desired. See *Figure 7.23*:

Figure 7.23 – New brush created from an image

After experimenting with our new Brush, we are ready to cover the next section to further understand how to navigate and export the new brush we have created.

Exporting the new brush

Go to the **Brush Tool** and click the white arrow button to export the new **Brush** as a `.abr` document. You can find the new Brush in your computer's `Downloads` folder. See *Figure 7.24*:

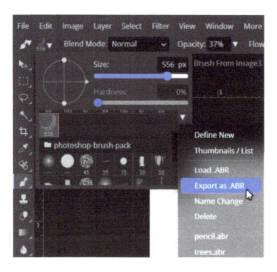

Figure 7.24 – Exporting the new brush as a .abr document

Make sure to rename the exported **Brush** .abr file to a name you can identify whenever you need to import it back into the **Brush** panel.

> **Important note**
>
> An important note about creating new Brushes: the new Brushes are only temporarily available. Once you close the Photopea program, the new Brushes will not save. You need to export the Brush as a .abr file. You have to import (load) them back in and Photopea will ask if you want to keep and load the Brushes permanently.

More examples of custom brush from images

Repeating the same process for the banana, I created a custom **Brush** using an image of clouds. I took my time exploring the Size, Angle, Spacing, and so on. Once I was happy with the results, I renamed the Custom Brush **MB Cloud Brush**. See *Figure 7.25*:

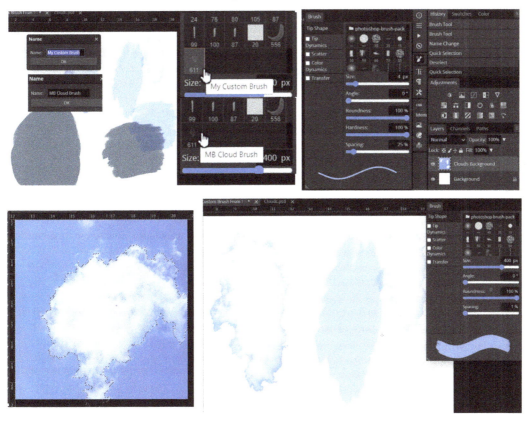

Figure 7.25 – Creating a brush from an image (of clouds)

I took the new cloud brush a step further by *changing* the **Angle** settings of the **MB Cloud Brush** for a different shape. See *Figure 7.26*:

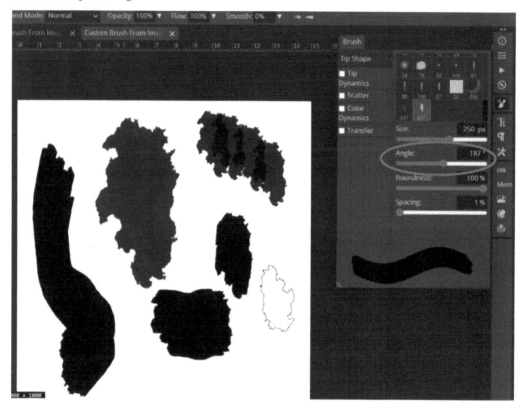

Figure 7.26 – Changing the Angle settings of the brush

Now that we've covered how to make brushes from images, we will now learn how to create a hand-drawn custom brush. Follow along with the steps in the next sub-section to do so.

Creating a hand-drawn custom brush

Creating hand-drawn custom brushes can be a lot of fun to come up with. It can be a good exercise to relax, build up your visual library, keep your brush library fresh, and further your understanding of brush Dynamics, and how they react with different Layer modes.

> **Important note**
> Creating hand-drawn custom brushes will require experimentation to come up with some interesting designs. Using layers and textures to experiment with drawing brush tips gives us more flexibility and possibilities to create unique brushes of our own.

Let's find out how to create a hand-drawn custom brush in this 11-step process:

1. To get started in this section, create a new document with a size of **1500 px x 1500 px.**

2. Create a new transparent Layer on top of the white default background, which will be used for the first concept.

3. Next, select the default **Hard Round Brush,** set its **Opacity** setting to **100 %**, and begin drawing with a solid black color.

4. After you create the first shape, make another transparent Layer and draw some other varied shapes, patterns, and lines that look like a good design or texture.

5. Continue building drawings and lines with different paintbrushes on separate Layers until you're satisfied with your Hand-Drawn Custom Brush shape. See *Figure 7.27*:

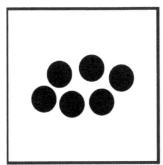 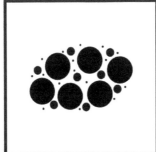 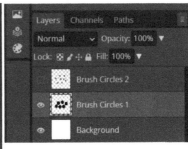

Figure 7.27 – Creating a hand-drawn brush

6. Once we have the desired shape, use the **Rectangle Select** Tool to select the new brush design. Next, go to the **Edit** menu, choose **Define New**, and select **Brush**, as seen in *Figure 7.28*.

The Brush is created instantly and activated for use.

7.   You can find the new brush in the **Brush** panel, and right-click over the new brush to rename it.

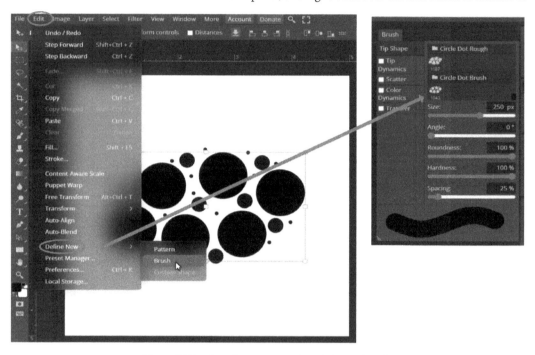

Figure 7.28 – Creating a custom hand-drawn brush

8. Next, test out the new Brush to see whether it will be a good style to add to the brush library. See *Figure 7.29*:

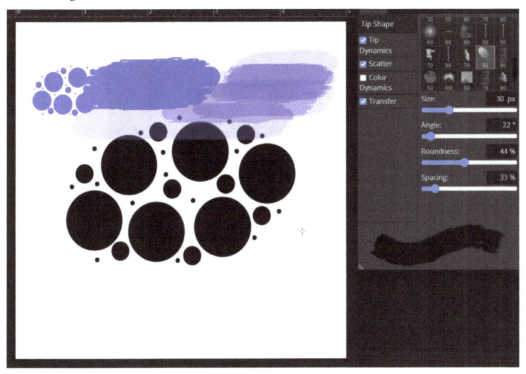

Figure 7.29 – Test out the new brush

9. I like this Brush and will export it as a .abr document.

10. Go to the **Brush Tool** click the white arrow button and select **Export as ABR**. The new brush will be saved as a .abr document. You can find the new brush in your computer's Downloads folder. See *Figure 7.30*:

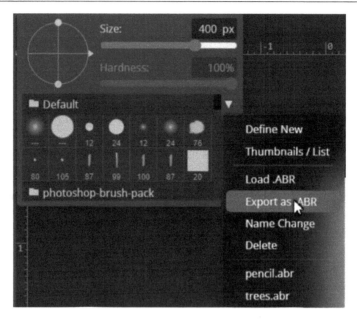

Figure 7.30 – Exporting a hand-drawn brush as a .abr document

11. Make sure to rename the exported brush .abr file to a name you can identify when you need to import it back into the **Brush** panel. See *Figure 7.31*:

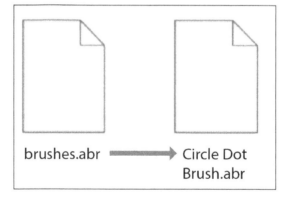

Figure 7.31 – Rename brushes.abr document

Important note

You will probably end up with several different brushes as you reveal and hide different drawing variations with the layers.

You can see examples of different Brushes I created from experimenting with the original Circle Dot Rough Brush I created, which I will discuss in the next section.

More examples of creating custom brushes

Here are more examples of creating custom brushes. I was able to add some additional marks and patterns and merge them with the initial custom hand-drawn brush, to create an altogether unique brush that I named **Circle Dot Rough Brush** and **Circle Scribble Brush**. See *Figure 7.32*:

Figure 7.32 – Circle Dot Rough Brush and Circle Scribble Brush

That wraps up this section on creating hand-drawn custom brushes in Photopea.

Brush management

Brush management is important for keeping your workflow organized. Imagine trying to access dozens or even hundreds, of brushes you've created without proper names or keywords; you may have made duplicate brushes by mistake, or you may have manually adjusted the brush dynamics to specific settings but forgot to save it. You would waste more time manually resetting the brush settings to see whether the brush is soft enough, hard enough, and has enough pressure or texture; all the while, you could have been working on your project.

These are some of the possible scenarios of mismanaging your brushes, and overall workspace for that matter. Let's look at some ways we can practice brush management, beginning with the Preset Manager.

Preset Manager

The **Preset Manager** is a great feature that enables you to manage and organize libraries for your brushes, gradients, patterns, layer styles, shapes, and contours. The Preset Manager may be useful for organizing some of your newly created brushes.

To access the Preset Manager, go to **Edit | Preset Manager**, as seen in *Figure 7.33*:

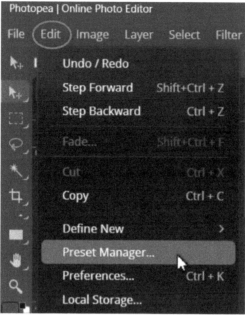

Figure 7.33 – Accessing the Preset Manager

1. This is the **Preset Manager** tab. Take a look at the preset **Brush** menu:

    *   **Load .ABR**: Load a brush from your local computer.

    *   **Export as .ABR**: Export a selected brush as a `.abr` file. You can use this to share your newly created brushes or use it to act as a backup and import them to load in the brush library permanently.

2. You can also review and use all of the **Preset Manager** categories (See *Figure 7.34*):

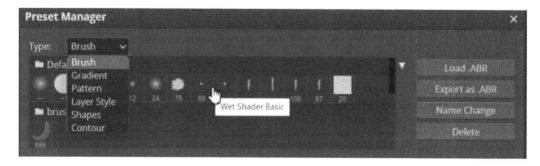

Figure 7.34 – The Preset Manager window

Now that we've touched on the Preset Manager for the brush Library, we can move on to deleting brushes.

Deleting brushes

Deleting brushes can be a little tricky if you're new to using the program. Deleting a brush in the **Brush** section of the **Presets Manager** window will only delete a brush *temporarily*. This can be annoying if you've created duplicate brushes and brush packs by accident.

To completely delete a brush or brush pack from Photopea, you need to delete it on your local computer drive.

To do this, go to the top menu under **Edit | Local Storage**, and delete the unwanted items permanently, as seen in *Figure 7.35*:

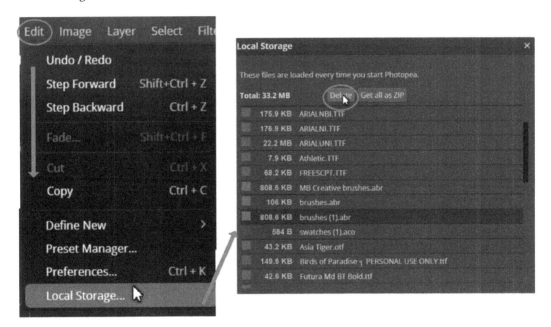

Figure 7.35 – Deleting brushes permanently

That covers this section on deleting brushes. Now, let's take a look at importing brushes from Photopea's free plugins and other library sources in the next section.

Importing Brushes from Photopea's free plugins and other library sources

You can check Photopea's free Plugins for free brushes. These resources are created by other artists, and new resources are updated and added regularly.

Go to **Window | Plugins| Brushes** (see *Figure 7.36*):

Figure 7.36 – Window | Plugins

Here are a variety of Brush Packs ready to install into Photopea's **Brush** library. See *Figure 7.37*:

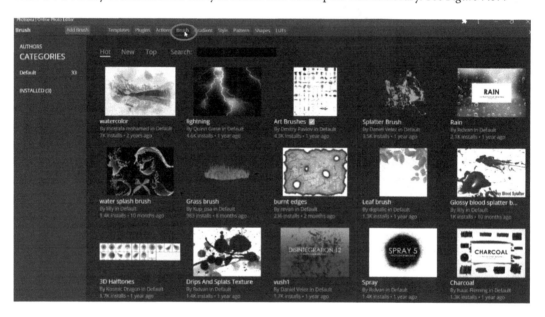

Figure 7.37 – Free brush packs in the Photopea Plugins menu

Select a brush pack (I chose the **Charcoal** brush pack), and click **Install**, as seen in *Figure 7.38*:

Figure 7.38 – Install brushes

You can find the newly installed brushes in the **Brush** panel, as seen in *Figure 7.39*:

Figure 7.39 – Testing the newly installed charcoal brushes

That wraps up testing newly installed charcoal brushes. Now, let's take a look at how to import brushes from other third-party resources.

Importing brushes from other third-party resources

Look for free or paid brush packs created for Photoshop, or other image-editing programs that have created brushes in the .abr format.

Here are some quick tips for importing, exporting, and renaming brushes.

Click on the white arrow located on the right side of the **Brush** panel to export, name change, load, delete, and so on. See *Figure 7.40*:

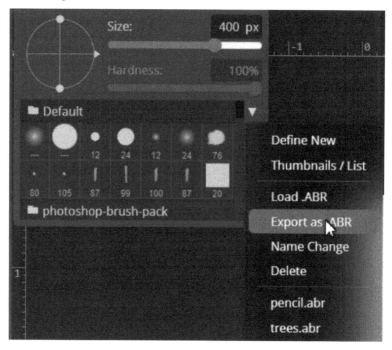

Figure 7.40 – Click the white arrow; reveals Export as .ABR, Load .ABR, and Define New options

You can right-click a brush to rename it. You can also load and delete brushes in the **Brush** panel. It will only delete the brushes temporarily. Remember, you have to delete the brushes under **Edit | Local Storage** to permanently remove them (see *Figure 7.41*):

 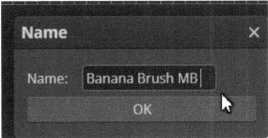

Figure 7.41 – Right-click the brush to rename it

That concludes the *Importing brushes from other third-party resources* section.

Summary

That covers this chapter on using and creating brushes in Photopea. We learned how to access the brushes and about Photopea's default brush library. We experimented with a number of default brushes to understand and see how each one reacts and marks differently on the canvas.

We also learned how to edit the default brushes and create a new brush by changing the brush settings and **Tip Dynamics** on the original brush.

In addition to working with the default brushes, we were able to create hand-drawn brushes of our own. By experimenting with different designs that we drew on separate layers, we were able to better experiment and see which designs worked best together to make a brush. To go a step further, we created brushes from photo images, using their shapes to create unique brushes.

Brush management is another important section we covered. We learned how to import and export brushes as .abr files, which gives us the ability to share brushes that we've created, as well as having a backup file we created, in case we lose the brushes due to a computer crash, for example. We also learned how to rename the brushes.

> Important note
>
> As we dive further into photo editing, we will learn how basic tools, such as the Eraser tool, Blur tool, and Smudge tool aid and assist the Brush and Pencil tool, with the ability to modify areas of an image's color, edges, texture, saturation, and so on. This will begin with the next chapter on photo retouching techniques.

In the next chapter, we will expand on what we've learned so far, and explore photo retouching techniques in Photopea.

8
Photo Retouching Techniques

In this chapter, we will break down the fundamentals of photo retouching. Photo retouching brings out the best in a person's features, looks, and style. It can help enhance images from photoshoots as well as for images used in publications, ads, prints, and products with a polished and professional look, by removing unwanted elements and distracting marks.

The basic tools (Brush, Clone Stamp, Eraser, Pencil, Layers, Masks, and so on) will aid in executing the retouching exercises. We will expand our understanding of smoothing out skin tones; removing unwanted marks and blemishes on skin; editing and shaping hair; reviving an old, worn photo; establishing the best possible exposure; and adjusting RAW images. This will enhance your knowledge and skills in photo retouching and working in Photopea.

For this chapter, we will cover the following topics:

- Touching up a portrait
- Adjusting and adding hair
- Restoring an old photo
- Adjusting an image in RAW format

Touching up a portrait

> **Important note**
> Before we start, I urge you to take your time. The program can lag at times, and if you move too fast, the tools may do unpredictable things, sometimes it may not save your progress right away, and so on.

People can be the most challenging subject to touch up in photo editing. There are endless variations of features that vary from one person to the next, including skin tone (such as freckles or scars), hair texture, eyes, and smile, which can appear more or less prominent under different lighting and backgrounds.

Understanding these things helps with touching up gradually and carefully so the end result will be a natural, soft look rather than a stiff, flat, harsh, or artificial computer-generated look.

Now, let's explore how to touch up a portrait step by step in the upcoming section.

Choosing and preparing a portrait to touch up

As you continue learning and building up your photo and image library for your projects, you will discover there is almost always going to be something in your image that needs to be edited or tweaked, such as adjusting the contrast, cropping out parts of a photo, or adding to a photo.

In this first example, I am going to edit the photo before getting started on touching up the face:

1.   On observing the image, I feel the need to crop out the figure on the left side with the crop tool. See *Figure 8.1*:

Figure 8.1 – Crop the figure on the left

Important note
Save a copy of the original photo in your document folder until you are finished and satisfied with your edits and final project.

2. Next, make a duplicate Layer of the cropped photo as a backup.

3. I feel like the edges of this person's hair are too close to the top edges of the photo. To fix this, I will scale down the portrait and add some of the blue sky on a separate layer. See *Figure 8.2*:

Figure 8.2 – Duplicate layer and scaled-down image

4. Select the **Eyedropper** tool to sample the blue sky color.

5.   Select **Soft Brush**, then set the brush's **Hardness** property to **0%** and **Opacity** to **20%**.

Paint in the area. See *Figure 8.3*:

Figure 8.3 – Select Eyedropper to sample the blue sky and paint in the area

I also feel that I could make the white cloud in the top-right corner less prominent around the hair by painting in some more blue sky. See *Figure 8.4*:

Figure 8.4 – Paint blue to make the cloud area more subtle

6. I decided to experiment further with the upper-right corner of the sky. I copied a small section of the cloud to the right of her ear with the lasso tool.

I moved it to the top corner and dropped the Layer **Fill** property down to **37%**. See *Figure 8.5*:

Figure 8.5 – Clone the lower cloud and move to the top

Let's see a before and after of the image with the changes to the clouds and sky side by side. I feel that making those changes to the top right of the sky improved the composition, with more emphasis on the lady. See *Figure 8.6*:

Figure 8.6 – Compare both backgrounds

Now that we've adjusted the background and cropped the image, let's touch up the lady's face and hair.

Touching up the face and hair

After taking a moment to observe the portrait, I will begin removing pimples and small dark spots by using the **Spot Healing Brush** tool.

To do this, select **Spot Healing Brush Tool** and set the **Hardness** property to **0%**. See *Figure 8.7*:

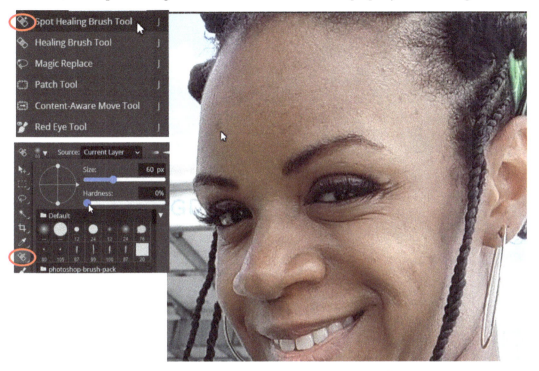

Figure 8.7 – Observe dark spots, then select Spot Healing Brush Tool

Keep the brush size similar to the size of the spots you would like to remove and paint short lines along the path of the spot.

Let's see a before and after of the face using **Spot Healing Brush**, in *Figure 8.8*:

Figure 8.8 – Removing dark spots with Spot Healing Brush

Now that we've covered removing blemishes and dark spots, let's move on to how we can soften lines and wrinkles around the eyes.

> **Important note**
> Avoid using **Spot Healing Brush** on the lines on the face around the smile and eyes, and also the eyebrows, lips, and so on, as it will ruin the pixels.

Softening lines and wrinkles around the eyes

Select **Healing Brush Tool** (not **Spot Healing Brush Tool**), and change the Brush's **Hardness** property to **0%**.

Sample the skin tone color near the eye by pressing the *Alt* key and *left mouse click*. Make sure the brush size is not too big, or you will paint in unwanted colors.

Begin painting over the wrinkles under the eyes with short brush strokes. Leave some hints of the lines to make it look natural. See *Figure 8.9*:

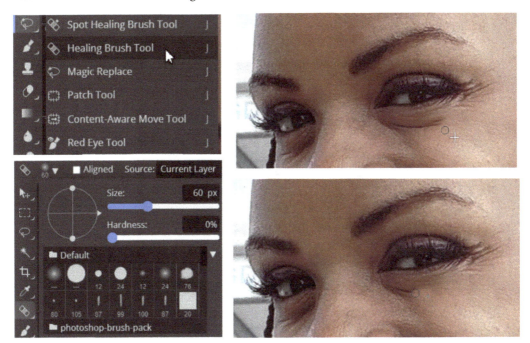

Figure 8.9 – Soften eye wrinkles with Healing Brush

You may need to sample skin color from other areas of the face to match areas where the colors differ too much in value. For example, I sampled the color underneath the eye on the right to touch up the dark color under the eye on the left to keep it from looking too dark and heavy compared to the right side. See *Figure 8.10*:

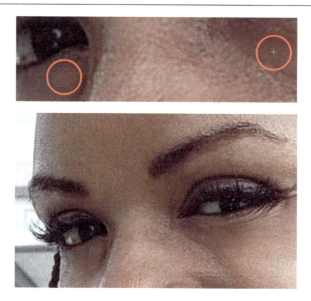

Figure 8.10 – Sampled right side of the eye to paint the left eye with Healing Brush

The eyes have a smoother appearance, and the wrinkles are removed, as seen in *Figure 8.11*:

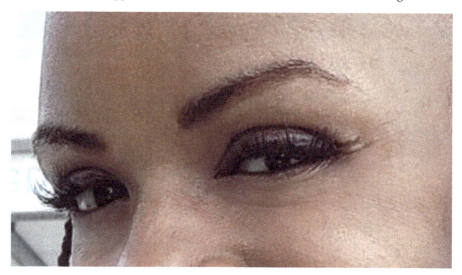

Figure 8.11 – Eye wrinkles removed

This particular image didn't need much smoothing on the face. Her skin texture was naturally smooth. Next, let's take a look at the image and observe the hair for any unneeded hair strands.

Adjusting and adding hair

I felt the image we retouched of the lady in the previous section could use some work around areas of the hair. We will remove some of the strands of hair, but not too much, to keep a natural look. We don't want the hair to look too perfect and evenly laid down. See *Figure 8.12*:

Figure 8.12 – Observing the hair to make touch-up adjustments

Now that we've observed the photo and considered what areas we might want to work on, let's work on cleaning up the hair in the next section.

Cleaning up the hair

As stated, hair can be tricky when you have thin layers that may get mixed into complex backgrounds for making selections. In this example, we will not be removing her from the background, just cleaning up some of the messy hair.

I've found that **Soft Round Brush**, alongside the **Clone** tool, is sufficient to cover up loose, messy hair.

I selected the **Color** picker to sample the blue sky color.

Set the brush's **Hardness** property to **0%** and **Opacity** to a value ranging from **20%** to **35%**, and paint over the hair using a separate Layer. See *Figure 8.13*:

Figure 8.13 – Color picker used to paint blue on a separate layer

I continued painting with a **Soft Round Brush**; you can see the result on the right side of the preceding figure. I needed to paint over parts of the hair, then redrew in some thin hair lines on another separate layer. Here, you can see where I painted out half of the longest hair curl and then redrew some of it to curl in. See *Figure 8.14*:

Figure 8.14 – Removing and adding hair with a Soft Round Brush

Let's hide the portrait to see what areas I painted on the separate layers, as seen in *Figure 8.15*:

Figure 8.15 – Areas I painted over

Although using masks is effective and important for managing complex sections and selections, not all situations require a mask to get similar results for subtle areas.

Let's make another comparison of the original image compared to the updated version with the hair removed and parts of the hair strands drawn in. See *Figure 8.16*:

Figure 8.16 – New comparison with hair and sky adjusted

That covers this section on cleaning up hair. Let's move on to the next section on how to restore an old photo.

> **Important note**
>
> *The Refine Edge Tool* section in *Chapter 5, Understanding Selection Fundamentals*, covers making hair selections.

Restoring an old photo

I remember seeing some of my old photos from kindergarten and junior high school that had scratches, stains, tears, and wear. For years, I wanted to restore them but didn't have the skillset or knowledge of how to go about doing so. As I learned about and gained more experience with image editing, I was able to restore them to near-perfect condition.

Another photo I was proud of restoring was one of my now late grandmother's mother on my late father's side. It was a small black and white **5" x 7"** photo that had a lot of scratches and spots and also some distortion that I was able to clean up and enlarge to a **16" x 20"** photo of my grandmother's mother. It was extremely gratifying to see the look of satisfaction on my grandmother's face when she saw the resulting photo that I proudly hung up in her front room.

Let's look at one of many approaches to restoring an old photo, beginning with removing spots and scratches. The photo in the upcoming section is of my grandmother on my mother's side of the family.

Removing spots and scratches

Thanks to image editing technology, old photos from earlier years and generations of family members can be scanned, restored, and saved in digital format. Some years back, I was able to restore a black and white photo of my grandmother's mother. It was a small **4"x 6"** worn and scratched photo that I fixed and enlarged to a framed **16 "x 20"** image. It gave me so much satisfaction to see the look on my grandmother's face when I handed it to her.

In the next example, I will demonstrate one of several ways you can restore an old photo.

Bringing your image into Photopea

To get started, locate your photo on your computer. You can drag your photo from your folder directly into the main area of the workstation, or go to the **File** menu and upload it from the folder in which the photo is stored. Rename and resave it as a PSD file. Then, carry out the following steps:

1. Check to make sure the image Layer has been converted into a raster image.
2. Make a duplicate Layer of the image.
3. Select the **Spot Healing Brush Tool** and set **Hardness** to **0%** (see *Figure 8.17*).

We will start eliminating the small marks and thin lines using **Spot Healing Brush**.

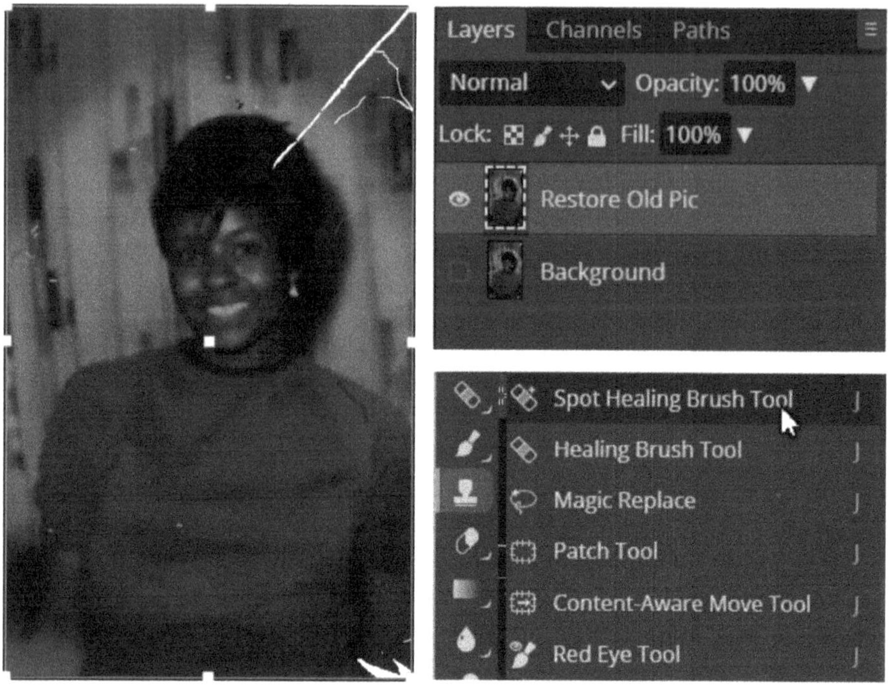

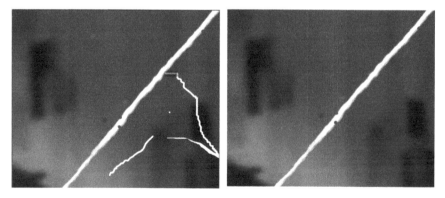

Figure 8.17 – Duplicate the Layer, then select Spot Healing Brush Tool

4. Set the Brush Size similar to the width of the scratch. This reduces the chance of the brush picking up other colors and objects in the background.

5. Begin making short brush strokes over the prominent areas first. In *Figure 8.18*, you can see **Spot Healing Brush** remove the thin white lines and small dots in the small area I cleaned up:

Figure 8.18 – Set brush size to the size of the scratch and make short brush strokes

6. For the wider marks, we can use the **Clone** tool or the **Healing Brush** tool. Also, if you still notice light brush marks from the **Healing Brush**, switch to the **Healing Brush** tool, press the *Alt* key to sample the area you are trying to match, release *Alt*, and brush over that area again (if needed). See *Figure 8.19*:

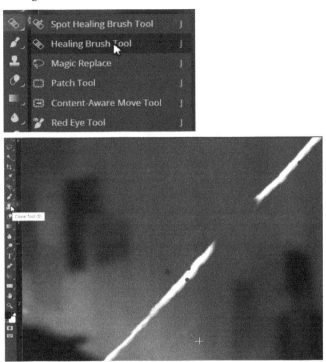

Figure 8.19 – Use Healing Brush to remove wider scratch marks

7. Continue to use the **Spot Healing Brush** on the face, hair, and shirt. See *Figure 8.20*:

Figure 8.20 – Use Spot Healing Brush to remove the scratches on the face

8.   Continue on to the shirt and arms using **Spot Healing Brush**. See *Figure 8.21*:

Figure 8.21 – Use Spot Healing Brush to remove the scratches on the shirt and arms

9.   As we make our way to the lower area of the photo, there are large spots with discoloration on the left and white scratches on the right side, by the forearms. It will be most efficient to use the **Crop** tool and crop off that area of the photo. See *Figure 8.22*:

Figure 8.22 – Crop out the discolored spot and scratches on the lower section

That ends this section on removing spots and scratches and using the Crop tool to remove sections of the picture that are not that important. Let's move on to the next section on color-correcting an image.

Color correcting an image

Learning to add color-correcting enchantments to an image will give your images a fresh and vibrant look. It can be refined into a work of art! Let's take a look at how to do that with the photo that we touched up:

1. Make a duplicate layer with the recent changes to the photo.

2. Select **New Adjustment Layer**, then select **Curves**, and adjust the image until satisfied with the adjustments. See *Figure 8.23*:

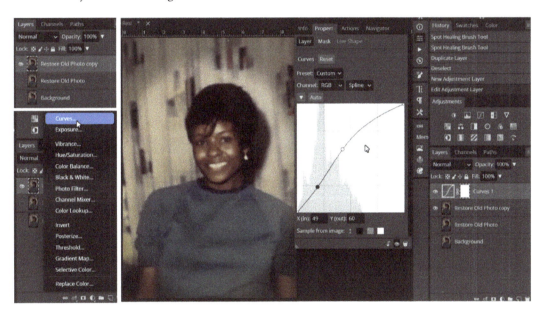

Figure 8.23 – Enhance image with Curves

3. One more adjustment I decided to make was lightening up the lower-left corner with the shadow. I used the **Clone** brush to sample a lighter area just above the shadow so it wouldn't stand out so much. See *Figure 8.24*:

Figure 8.24 – Remove the lower-left shadow with the Clone tool

---

Important note

Also, for this particular photo, I could have used the **Color** picker to sample an area of the photo and paint in the matching colors with the Soft Brush tool. The reason for this is that the photo already has a slightly softened painted look to it. Don't be scared to test out the tools; it will only heighten your skills and understanding of the tools and program.

Since this is an older image, we may not be able to add any more adjustments to improve it. Something else you can experiment with is converting your image to black and white, or some other monochromatic or burnt sienna-type looks.

That sums up one way we can restore an old photo. Let's begin the next section on adjusting an image in RAW format.

Adjusting an image in RAW format

If you are a photographer, you may not want to compress all of your shots to JPEG. Instead, you save the photos as RAW images in your camera to save all of the information in that image, for example, retaining colors or exposure levels that may get lost in a compressed JPEG.

If you decide you want to display it on the web and/or make adjustments to the image, you would need an image-editing program that can open RAW files, make adjustments, and export them as other formats, such as JPEG, PSD, and PNG, to display on the web or get printed copies of the image.

Photopea can open RAW files and also has a **Raw Camera Filter** mode to make some essential adjustments. It doesn't have as many features as Photoshop or the Lightroom application, but you can still achieve similar results if you use it along with filters, Layers, and image adjustments. You could also download a free alternative such as Darktable to give you more editing power for RAW files.

Now, let's get started and see how we can adjust an image with the **Camera Raw Filter** in Photopea:

1. Open the image and duplicate the image Layer.

2. Next, go to **Filter | Camera Raw Filter**. See *Figure 8.25*:

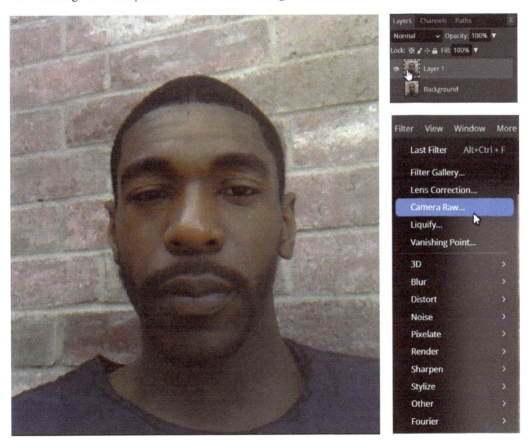

Figure 8.25 – Duplicate image layer and launch Camera RAW Filter

3. When you select **Camera Raw Filter**, a new window panel opens with six image adjustments you can change to get a variety of results for your image: **Temperature**, **Tint**, **Exposure**, **Contrast**, **Vibrance**, and **Saturation**. See *Figure 8.26*:

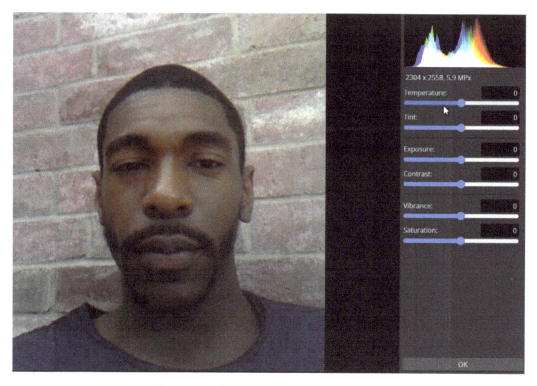

Figure 8.26 – Camera RAW Filter panel window

4. I feel this image could use some adjustments of **Temperature** and **Saturation** at first glance. I will explore adjustments to try and color-correct as best as possible with **Raw Camera Filter**.

5. I adjusted the **Temperature** from **0** to **-36**. This helps the man stand out from the reddish brick wall behind him. See *Figure 8.27*:

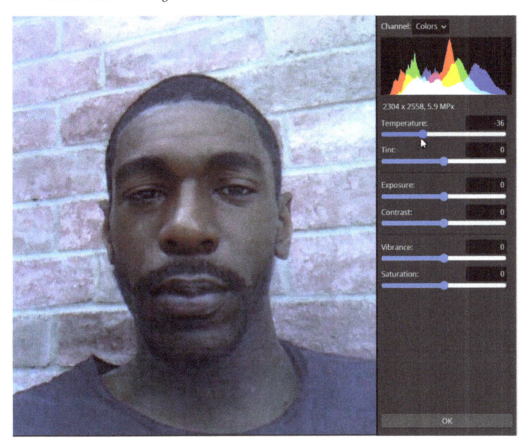

Figure 8.27 – Adjust Temperature with the Raw Camera Filter

6. After trying a variety of settings, I settled with the Temperature adjustments made in *Figure 8.27*. I was unable to remove the over-saturated red on the nose and left cheek area in **Camera Raw**.

7. The next step is to click **OK** in **Camera Raw** to save that version and create a new adjustment layer for **Hue/Saturation**:

8. In the **Hue/Saturation** panel, I selected **Red** for **Range** and changed **Saturation** to **-91** (see *Figure 8.28*). That helped balance the skin tone and color balance of the face nicely.

9. If I wanted to try any more variations with the **Camera Raw Filter**, I would need to merge and flatten the **Hue/Saturation** layer with the image of the man to open the **RAW Filter** panel.

10. I feel the results turned out well with the **Camera Raw Filter**. See *Figure 8.28*:

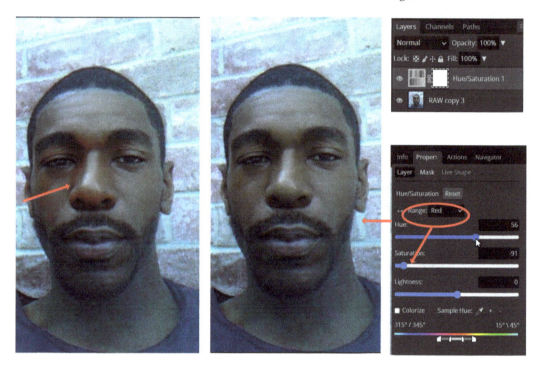

Figure 8.28 – Making further adjustments with the Hue/Saturation adjustment layer

Let's see what else we can do with the **RAW Camera Filter** for fun. See *Figure 8.29*:

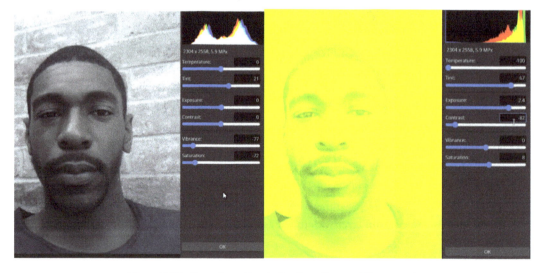

Figure 8.29 – Experimenting in the Raw Camera Filter panel

After exploring and experimenting with the Camera Raw filter, we have a better understanding of how to make adjustments to images shot in RAW format. That sums up this section.

Summary

This marks the end of this chapter on the fundamentals of photo retouching techniques in Photopea. We covered a variety of techniques for different use cases, such as touching up a portrait, prepping a photo before applying touch up work, restoring an old photo, and learned how to use and apply adjustments with the Camera Raw Filter, similar with limitations to making adjustments with RAW files in programs like Adobe's Light Room.

Exploring techniques further, we learned how to remove blemishes, pimples, and marks with the Spot Healing Brush, Healing Brush, and Clone tools. We used the Brush tool to paint over subtle areas of the hair, sky, and clouds, without using masks.

In addition, we made new Layer adjustments to enhance the color, contrast, and saturation, and when and how to apply color-correcting and color temperature to an image.

In closing, we covered a lot in this chapter, which has prepared us for learning advanced image compositing techniques in the upcoming chapter.

9

Exploring Advanced Image Compositing Techniques

In this chapter, we will take all the things we've learned to another level. We will make adjustments to images through lighting, retouching, selecting, combining, and arranging various object elements and backgrounds into a single believable composition, with the power of image compositing.

By the end of this chapter, you will have gained an understanding of what Image Compositing is and know how to execute an Image Composite from start to finish with confidence.

For this chapter, we will cover the following topics:

- What are Image Composites?
- Preparation for creating Composites
- Applying skills to an Image Composite – Lunch on Table
- Applying skills to an Image Composite – Man Walking

What are Image Composites?

Image Composites are accomplished by combining two or more images, such as a photograph, along with 3d assets and design elements, to create a new image or composition. Great examples of Image Composites are *movie posters*, *advertisements*, *collages*, and *illustrations*. They can capture and express different modes and movements and depict different eras or unique genres, such as *fantasy*, *sci-fi*, and *horror*.

Image Composites can also be used to help concept artists and illustrators devise ideas without having to depend on working out ideas from scratch.

With image compositing, artists can develop ideas for a new creature or character for **character design**, by swapping out different fashion and clothing, anatomical and facial features from many different photos of people, cultures, and customs.

A similar approach can be applied to **creature design** – using and swapping images of animals, insects, and creature's features and anatomical makeup to create a unique-looking creature.

In addition, **environment design** can also be achieved with Image Compositing by combining different architectural buildings and landscape scenery.

Once an Image Composite is created, artists can create sketches and models from them as the basis of studies for finished designs.

Let's explore and experiment with image compositing in the upcoming sections.

Preparation for creating composites

Working in the industry as an illustrator and graphic designer over the years has given me unmeasurable experience and ways to prepare and approach ideas. With that in mind, here are a few things to remember when creating Image Composites:

- **Sketching out ideas**: Keeping a sketchbook/journal close by to brainstorm ideas, write descriptions and notes, or explain the mood, emotion, or story you want your image composite to tell or portray can help if you feel stuck, or can't seem to come up with something good right away.

  At times, creating rough sketches can give you an idea of the types of photo references you require to create an Image Composite. This may help you stay focused and organized, and it may also save you some time if the ideas created for the sketches (i.e., thumbnails) don't turn out so well.

- **Setting up a photo shoot for your Image Composites**: Since great photos and images are an integral key in creating Image Composites, it's great to have good images and photos to work with. Photopea provides some (limited) stock images that are ready to work with, located under **Window** | **More** | **Plugins**.

  You can also check out different open source stock image sites such as the following:

  - `www.pexels.com`
  - `www.unsplash.com`
  - `www.freeimages.com`

Searching these sites can sometimes be overwhelming and time-consuming, with so many images to choose from. This is why, sometimes, the quickest way to gather the best photo resources may require you to set up a photo shoot of your own.

In the next section, I will provide an example of a project that I set up for a photoshoot, called *Lunch on Table*.

Applying skills to an Image Composite – Lunch on Table

In this example, I shot several shots of a plate of food. A couple of the photos were taken at different times because I was undecided on what I wanted to include in the arrangement of the table setup.

I provided these photos in the resources folder for you to follow along; as I think it will be a good exercise to re-enforce some of the skills we've learned thus far – *making selections*, *creating masks*, *making adjustment layers*, and *refining edges* – allowing us to combine photos to make an Image Composite, which is the overall lesson of this chapter (see *Figure 9.1*):

Figure 9.1 – Shooting Your Photos for Image Composites

Now, let's see how we can use these different shots to create an Image Composite by making *selections*, *adjustments*, and *re-arranging* them to create a new composition:

1. At the top-left of *Figure 9.1*, in the first photo, is some food on an orange plate and a multicolor blue mat underneath. We want to make a *quick selection* around the *orange plate*.

2. Go to the **Select** menu | **Modify** | **Contract** (**Contract** will unselect some of the pixels that you don't need). In this case, it will remove some of the pixels that went beyond the plate area (see *Figure 9.2*):

Figure 9.2 – Use the Quick Selection tool to select the plate

3.   While the *plate* is selected, go to the **Edit** menu and select **Copy**, or press *Ctrl + C*, to make a copy of the plate and then *Ctrl + V* to paste a new copy of the plate of food (see *Figure 9.3*):

Figure 9.3 – While selected, copy and paste the plate of food

After making a copy of the plate of food, I decided to finish the project in a new document by pasting the selected plate of food into it.

4.   To do this, go to **File | New** and create a new document, sized **11.5" x 8.5"**.

5.   Next, press *Ctrl + V* to paste the **Orange Plate Food on Mat** Layer onto the transparent background of the new document (see *Figure 9.4*).

> **Important note**
>
> This will enable us to replace the *table* and *table mat* with a different-styled table and or table mat, respectively, if desired.

Figure 9.4 – Paste the plate of food onto a new document

6. Now that we've added the plate of food to a new document, let's add a different texture underneath the plate to represent the table.

7. You can drag the gray textured image from the resources folder directly into your document, (or use your own texture), and then *resize* it to fit in the entire background Layer (see *Figure 9.5*):

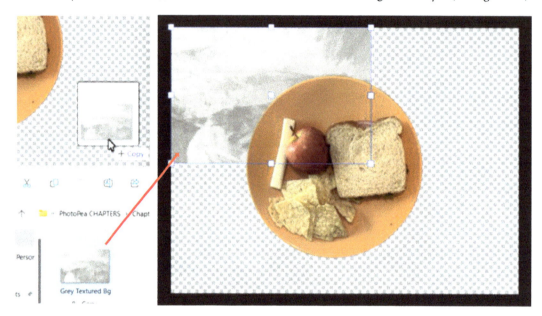

Figure 9.5 – Drag a textured image into the document and resize it

After resizing the textured table, we can add a drop shadow along the edge of the orange plate.

8. Double-click on the **Orange Plate Food** layer to activate the **Layer Style** panel; alternatively, *click* the **Layer Style** button, located at the bottom of the **Layers** panel. This will give the plate a sense of weight and dimension. Otherwise, the orange plate food will look like a cutout, floating on top of the textured background.

9. Next, experiment with the **Drop Shadow** settings for **Angle**, **Distance**, **Spread**, and so on (see *Figure 9.6*):

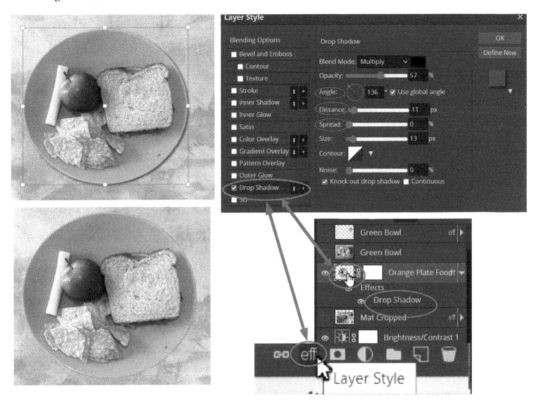

Figure 9.6 – Add a drop shadow to the plate

10. Next, *select* the **Rectangle Select** tool to make a rectangular selection around the blue mat.

11. Next, press *Ctrl + C* to make a copy of the blue mat, and place it underneath the **Orange Plate Food** layer. The original shot of the mat was skewed, due to my not positioning the camera straight (see *Figure 9.7*):

Figure 9.7 – Adding the table mat

12. Next, add a drop shadow around the *table mat*. Repeat the process for the plate of food (see *Figure 9.8*).

Important note

We can always go back and edit the Drop Shadows and other *Layer effects* as we continue to add images and make adjustments.

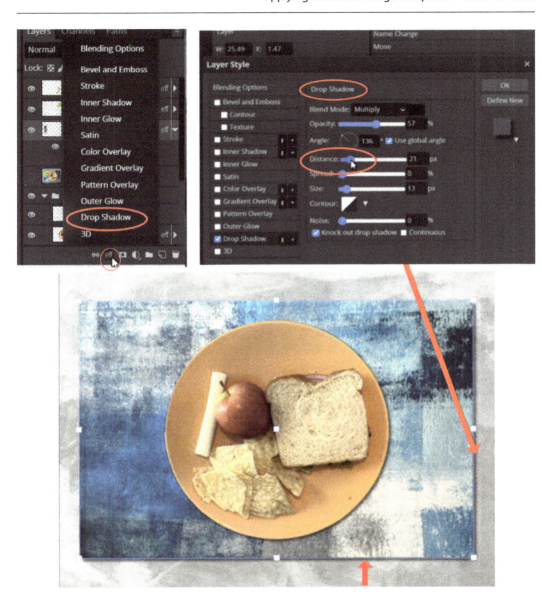

Figure 9.8 – Add a drop shadow around the table mat

13. Now that we've added the **Drop Shadow**, click the **New Adjustment Layer** at the bottom of the Layers panel, and select the **Brightness/Contrast** Properties panel to darken **Grey Textured Bg** (see *Figure 9.9*):

Figure 9.9 – Adjust the brightness and contrast for the table texture

14. Next, trace around the green bowl using the **Pen** Tool to make a selection around the bowl.

15. Choose a color, such as orange, and make the **Stroke** size thin.

This will make it easier to see where you are placing the anchor points, and you can stay close to the green bowl, avoiding the table being in the selection. Click and drag as you place each node. The more space you have in between Nodes, the smoother your path (outline) will be (see *Figure 9.10*):

Figure 9.10 – Trace the bowl image with the Pen tool

16. After you've closed the path with the last node, change the pen mode from **Shape** to **Path**, and then click **Make Selection** to make the selection active.

17. Duplicate the **Green Bowl** layer, and press *Ctrl + J* on the Layer to remove it from the background.

18. Copy and paste the green bowl with a transparent background into our *Lunch on Table* project (see *Figure 9.11*):

Figure 9.11 – Turn the Pen tool shape into a path and make a selection

19. After making the selection, *copy* and *paste* the Green Bowl onto the document we are working on with the **Orange Plate Food Layer**, and add a drop shadow to the edge of the bowl.

    Now, we can import the banana we used from *Chapter 5, Understanding Selection Fundamentals*, into our project.

20. Before we do so, let's go to the **Image | Adjustments | Curves**.

    Make the necessary adjustments to add contrast to the banana, since the original shot involved more light than the **Orange Plate Food** shot (see *Figure 9.12*):

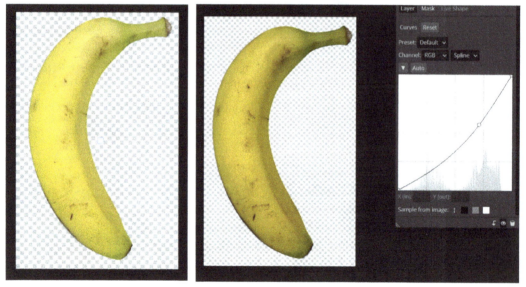

Figure 9.12 – Open the banana from Chapter 5, Understanding
Selection Fundamentals and go to Image | Adjustments

21. Copy and paste the banana into our *project* and add a Drop Shadow from the **Layer Style** window (see *Figure 9.13*):

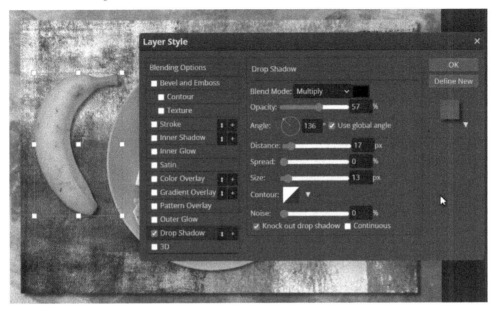

Figure 9.13 – Copy and paste the banana and add a drop shadow

22. Next, open the photo with the silverware.

23. Rasterize the layer, and then duplicate it (see *Figure 9.14*):

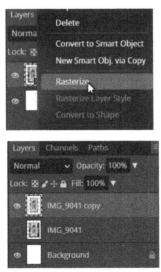

Figure 9.14 – Open the silverware photo and rasterize and duplicate it

24. Next, use the **Quick Selection** tool to select the silverware on the white napkin. While the silverware is selected, click the **Refine Edge** button to open the **Refine Edge** window (see *Figure 9.15*):

Figure 9.15 – Select the silverware and open the Refine Edge window

25. Select the *Brush* on the top-left side, and make sure the gray button is active to begin softening and refining the edges of the napkin. This removes unwanted pixels and color along its edges (see *Figure 9.16*):

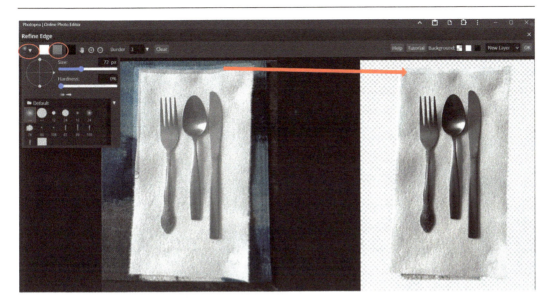

Figure 9.16 – Select the brush on the left side to paint along the edges of the napkin

26. After refining the edges of the napkin, click **OK** to make the changes and close the **Refine Edge** window.

27. Add a layer, and fill it with a solid color to see how well the edges of the napkin have been refined (see *Figure 9.17*):

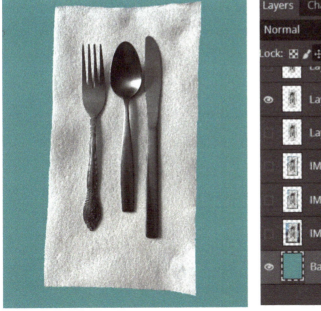

Figure 9.17 – Add a color background to inspect the refined selection

28. After viewing the selection of the napkin, we can place it into the project and add a Drop Shadow to it (see *Figure 9.18*):

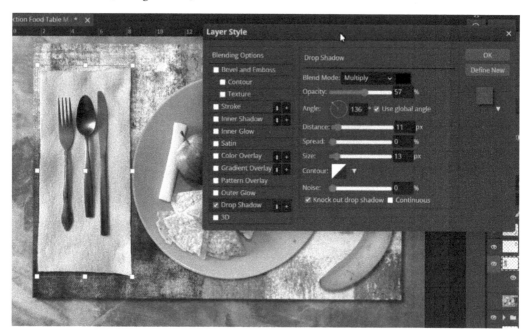

Figure 9.18 – Place the napkin silverware and add a drop shadow

29. Observing the orange plate, I felt the orange was dull, so I increased the **Vibrance** setting to **34** (see *Figure 9.19*):

Figure 9.19 – Increase the vibrancy of the orange plate

30. Next, I created an adjustment Layer to adjust the Curves on the overall image and layers (see *Figure 9.20*):

Figure 9.20 – Adjust the Curves on the Overall Image

31. We can now take a look and see whether there are any other effects or enhancements that we can apply to this *Lunch on Table* project (see *Figure 9.21*):

Figure 9.21 – Observe the image and make any necessary adjustments

This concludes the *Applying Skills to an Image Composite – Lunch on Table* section.

In the upcoming section, let's look at another example of creating an Image Composite using figures and landscapes.

Applying skills to an Image Composite – Man Walking

Important things to remember about Image Composites

There is no one specific way to create an Image Composite. One composite may only need a few steps, whereas another might take 10 steps. It can be challenging to know when the composite is finished because you can constantly make tweaks or experiment with different moods on one image – for example, making the *Man Walking* image sunnier, happier, darker, or scarier. They would require different adjustments. The options are limitless.

Now, let's get started with the Image Composite.

The first thing to do is find some images, or you can shoot your own images that are similar in lighting, perspective, and time of day, or have a similar lighting setup with spotlights and so on.

I know I mentioned that it's best to have images with similar lighting and saturation values, but I was pressed for time to complete this section, and we will make adjustments manually to match the lighting and saturation as closely as we can.

I searched through shots and put together a mood board of photos I've taken on vacations that I felt would work for this section.

The shot of the Man Walking was taken in the Dominican Republic, and both mountain scenes were shot in Las Vegas. All three mountain scenes were taken at similar angles with the camera so we could make the images flow together.

The Man Walking has more saturated colors, but I can reduce the saturation with **Layer Adjustments**. The Raptures were taken in Atlanta, GA, and the Iguana in Florida was taken on a sunny day, but it may still work using **Layer Adjustments**. Let's see how it turns out (see *Figure 9.22*):

Figure 9.22 – The photo references mood board

I decided to make the horizontal Las Vegas mountains picture at the bottom left of *Figure 9.22* the main landscape to build the Image Composite.

Now that we've decided on the main image for the background, let's import the figure of the man walking into the main picture using the selection tools.

1. First, I decided to select the Fig Walking in the top left with the **Pen** tool to get the cleanest edges without picking up traces of the background (see *Figure 9.22*).

2. Make sure to make a *duplicate Layer* of the figure walking as a backup and hide the layer.

3. Select the **Pen** tool on the **Toolbox**, or press *P* to activate the **Pen** tool.

4. Choose a bright color (*green*) as the stroke color, and make the point size **.06** to easily draw along the path of the figure and remove it from the background cleanly (see *Figure 9.23*):

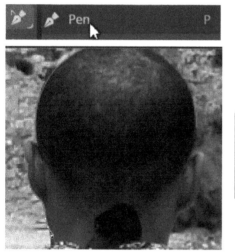

Figure 9.23 – Make a selection using the Pen tool

5. When you begin adding *anchor points*, a **Shape** Layer will automatically appear over the original image.

It's best to provide some space in between your anchor points so that you can click and slightly drag the mouse to make smooth curves when using the **Pen** tool (see *Figure 9.24*):

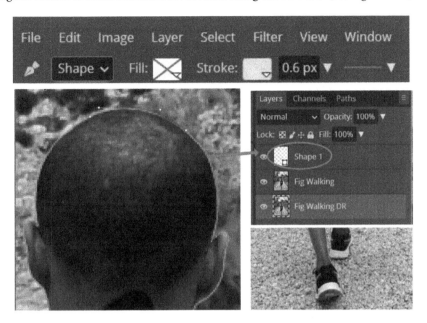

Figure 9.24 – Click and drag the anchor points with the Pen tool

Important note

If you need to edit (i.e., convert) the anchor point to a straight line or remove a point, you can press the *P* key or click on the **Pen** tool in the **Toolbox** to activate the **Add Anchor Point Pen** or **Delete Anchor Point Pen** tool.

6. Once you've finished tracing the Fig Walking with the **Pen** tool, switch the **Pen** mode from **Shape** to **Path**, and then click **Make Selection** to activate the selection (see *Figure 9.25*):

Figure 9.25 – Switch Pen mode to Path and click Make Selection

7. When the selection is activated, click on the Layer with the **Fig Walking 1** and press *Ctrl + J* to remove the figure from the background (see *Figure 9.26*):

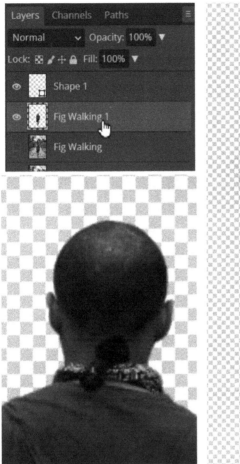 

Figure 9.26 – Press Ctrl + J to remove the background

8. Next, create a *new Layer*. Go to **Edit Menu | Fill Color**.

9. Choose a bright fill color and place it under **Fig Walking 1**, ensuring that it's clean and doesn't include any of the background (see *Figure 9.27*):

  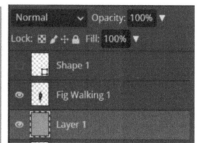

Figure 9.27 – The new layer fill color replacing the unwanted background color

---

Important note

Now that the figure is on a transparent background, you can always go back and make an active selection on him and the other images we removed from the background by selecting the layer, pressing and holding *Ctrl*, and double-clicking the layer.

---

10. With **Fig Walking 1** still selected, go to the **Select** menu | **Modify** | **Add Border** to add a **1 px** border around the figure.

11. Next, soften the edges around the **Fig Walking 1** Layer (the man walking), using the **Gaussian Blur** filter.

12. Adjust the **Gaussian Blur** setting to **2.4 px** (see *Figure 9.28*):

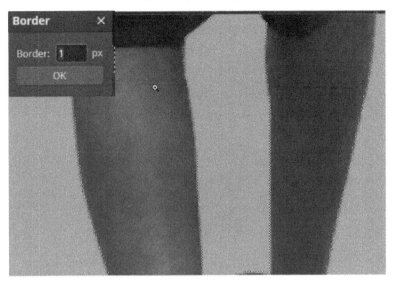

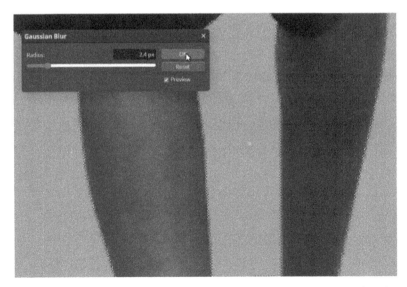

Figure 9.28 – Add a border around the selection and add a Gaussian blur to soften the edges

13. Next, I observed the selection again, and I noticed that part of the background was visible in the loophole of a shoestring and in between the left arm and the torso. To fix that, *deselect* the **Fig Walking 1** figure by pressing *Ctrl + D*.

14. Next, I used the **Eraser** tool and erased the background color. You can also try the **Magic Eraser Tool**; it may be an easier option to remove the background (see *Figure 9.29*):

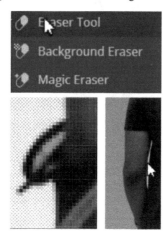

Figure 9.29 – Use the Eraser tool to erase the tiny areas of background color

Now that we've separated the figure from the background, we can repeat the same steps to remove the Iguana from the background using the **Pen** tool (see *Figure 9.30*):

1. Make a selection of the Iguana.

2. Make sure to make a *duplicate Layer*.

3. Repeat the same process, using the **Pen** tool to create a selection around the iguana.

4. Press *Ctrl + J* to remove the Iguana from the background.

5. With the Iguana still selected, go to the **Selection** menu and convert the selection around the Iguana to a **Border** selection of **2 px**.

6. Next, soften the edges around the Iguana using the **Gaussian Blur** filter (see *Figure 9.30*):

Figure 9.30 – Use the Pen tool to make a selection around the iguana and Gaussian blur the edges

7. Make a separate selection and Layer for the Iguana's Shadow so that we have more control over adjusting it (see *Figure 9.31*):

Figure 9.31 – Make a selection around the iguana's shadow

8. Now that we've removed the Iguana from the background, we can remove the Rapture from the sky.

9. Use the **Pen** tool to make a selection around the Rapture the same way we did for the Man Walking and the Iguana (see *Figure 9.32*):

Figure 9.32 – Use the Pen tool to make a selection around the rapture

Next, we will make a *new Layer* beneath the *Rapture Layer*:

1. Go to **Edit | Fill… | Color**, select **Black**, and fill in the color of the Rapture selection on the new layer.

2. Go to **Filter** on the top menu and select **Gaussian Blur**.

3. Adjust the **Radius** setting to **10 px** (see *Figure 9.33*):

Figure 9.33 – Fill the Rapture with black and Gaussian Blur the edges

4. Next, paste the Fig Walking, Iguana, and Rapture onto the horizontal photo of the Las Vegas mountains at the bottom left of the *Figure 9.22* reference sheet.

5. I arranged and placed the figures onto the landscape until I was satisfied with the composition arrangement (see *Figure 9.34*).

6. The Fig Walking, Rapture, and Iguana selections look good on the landscape, but they will need adjustments as we continue working on the Composite as a whole.

Figure 9.34 – Paste the figure walking, rapture, and iguana onto the Las Vegas photo

> **Important note**
> I renamed the Fig Walking Layer to Man Walking Layer.

7. Next, create a black and white *Adjustment Layer* to see how well the values (the darkest darks and brightest whites) and contrast balance out between the man walking, the Iguana, and the landscape scene (see *Figure 9.35*):

Figure 9.35 – Create a black and white adjustment layer to see whether the contrast is good

> **Important note**
>
> The goal is to create and balance similar lighting and values for the different images, while also keeping in mind that the man walking will be the main point of focus. He will need something to make him stand out a little bit more than the other objects in the Image Composite we're gradually building.

8. Next, *click* on the **Man Walking** Layer, create a **Curves** adjustment layer, and add a clipping mask to it. Repeat these steps for the iguana. (You will see a small arrow appear to the left of the **Curves** Layer, pointing down to the man walking.)

I made the Iguana slightly darker and slightly toned down the intensity of the Man Walking. You can see the placement of the points on the *Curves adjustment* diagrams (see *Figure 9.36*):

Figure 9.36 – Use the Curves adjustment layers and clipping masks to make adjustments

9. Next, create a **Curves** adjustment Layer for the Las Vegas background, followed by a clipping mask to make adjustments (see *Figure 9.37*).

I slightly darkened the Vegas background to compliment the dark values of the man walking. The photo's going in the right direction, gaining a mysterious and gloomy look.

Figure 9.37 – Create a Curves adjustment Layer for the Las Vegas photo

10. After making the adjustments, let's hide the *black and white Adjustment Layer* to view the *enhancements* of the Image Composite in color (see *Figure 9.38*):

Figure 9.38 – Hide the black and white adjustment layer to view the changes

11. There are a couple of things we can do next. We can tone down some of the harsh brights on the Iguana because we don't want it to be the focal point of the composition. We want to reduce some of the color saturation of the man walking so that he looks like he's in the same scene as the cloudy background. Colors aren't as bright and intense when the sky is filled with grey clouds.

Let's do all that now:

1. Let's start with the Man Walking. Create a **Hue/Saturation** adjustment Layer and add another Clipping Mask to the Man Walking.

2. Under the **Hue/Saturation** range, change **Master** to **Blue** for the man's bag.

3. Reduce **Saturation** to **-43** and **Lightness** to **-13**.

4. Next, change the **Range** to **Red** for the color saturation of the Man Walking's skin tone.

5. Reduce **Saturation** to **-10** and **Lightness** to **-10** (see *Figure 9.39*):

Figure 9.39 – Reduce the blue color saturation on the bag

Important note

After reducing the hue color saturation, I feel we need to tone down the bright values on the iguana; we need to do something different, so we will wait and revisit the iguana later.

I felt the overall Image Composite needs to have more trees and landscape foliage to help fill in the large empty areas of the composite.

Adding more trees and bushes to the Image Composite

I was halfway through this project and felt it needed more elements to add more depth and fullness to the composition. Adding more trees and foliage will create a more three-dimensional appearance, with overlapping tree elements on the flat landscape.

To begin the process, I made rough selections around the Desert Tree and Bushes (see *Figure 9.40*):

Figure 9.40 – Add more trees and bushes to the Image Composition

I then copied and pasted them onto the Image Composite. This enabled me to experiment with the placement and size of the plants quickly (see *Figure 9.41*):

Figure 9.41 – Roughly place the tree and bushes onto the Las Vegas Mountain photo

Once I was satisfied with the placement, I began making a selection around the Desert Tree and Bushes using the **Pen** Tool in a separate document.

I repeated the process using the **Pen** Tool for the Tree and the Bushes, and will place them back in the Image Composite for the next Layer adjustment (see *Figure 9.42*):

Figure 9.42 – Use the Pen tool to make clean selections on the buses and tree

Now that we've made selections around the Tree and Bushes, we can move on to the next section, *Adjusting the iguana's skin tone*.

Adjusting the iguana's skin tone

It's time to tone down the harsh highlights of the Iguana and add some brown to its skin so that it unifies with the shades of brown in the landscape's grass, rocks, and mountains.

1. Click on the **Iguana 1** layer, add a new **Hue/Saturation** adjustment layer under the **Curves** adjustment Layer of the Iguana, and then add a Clipping Mask.

2. Select **Green** under **Range**.

3. Set **Hue** to **-92**, **Saturation** to **-43**, and **Lightness** to **-6**.

    This setting unifies the Iguana with the foliage on the ground, the mountains, and the figure, giving the scene a natural look, and a sense of realism (see *Figure 9.43*):

Figure 9.43 – Tone down the harsh lightning with the Hue/Saturation and Curves layers

Now that we've toned down some of the hue/saturation and harsh lighting, let's take another step in reducing the harsh lighting by adding some overlays to color-match the figures, making them look more natural.

Working with absorbed light and color-matching

The next thing we need to do is add a green overlay to tone down the harsh lighting on the Iguana:

1. Click on the **Iguana** Layer, hold *Ctrl*, and double-click the mouse to make a selection around the iguana.

2. Select the color picker and sample the olive-green color of the Iguana. It's a neutral value that we can use to balance the harsh saturated colors (see *Figure 9.44*):

Figure 9.44 – Select the Eye Dropper tool to sample the iguana's green skin

3. *Ctrl* + double-click on the **Iguana**.

4. Create a new Layer, and then create a new **Fill Color Mask**, and title it *Color Fill 1*.

    Fill the Layer with the olive-green color, and you will then have a silhouette Mask of the Iguana (see *Figure 9.45*):

Figure 9.45 – Create a new Color Fill layer for the iguana

5. Switch the **Layer** mode to **Darken** and change **Opacity** to **55%**.

6. Select the **Soft Brush** tool and set **Hardness** to **0** by typing it in.

7. Change the foreground color to white and begin painting out the harsh colors. (switch the foreground color to black if you need to paint the original sections back in – see *Figure 9.46*):

Figure 9.46 – Change the opacity of the silhouette and paint out the green

8. Next, add another **Hue/Saturation** adjustment Layer, underneath the previous one, to the Iguana (Do not add a clipping mask.)

9. Select the **Red** range and slide the **Hue** property to **-56**.

   This will add more browns and reds to the Iguana and the foliage on the landscape (which is ironic since Iguanas naturally change their color for camouflage).

10. Then, add a **Hue/Saturation** Layer to the Las Vegas background with a Clipping Mask.

11. Adjust the **Green** range of **Hue** to **36** and **Lightness** to **33** (see *Figure 9.47*):

Figure 9.47 – Create Hue/Saturation adjustment Layers and add more brown

Take a moment to look over the adjustments made to the **Iguana** one more time.

12. Duplicate the olive-green color overlay to tone down the lighting.

13. Select the **Hard Brush Round**, this time to brush out the green (but leave some green on **Iguana** where the harsh lighting or brightest highlights are casting on it).

14. Now, add some vibrance to the Las Vegas background. Go to **Layer | New Adjustment Layer | Vibrance**.

15. Change the **Vibrance** property to **60** and **Saturation** to **-2** (see *Figure 9.48*):

Figure 9.48 – Duplicate the olive-green color overlay and add a Vibrance adjustment layer

Now that we've toned down The Man Walking and Iguana fairly well, and added some vibrance to the background, let's return to the man walking to focus on the smaller areas that stand out too much, that should be more subtle. For example, the white at the opening of the blue bag, the shoes, and so on.

Looking at the man walking at first glance, we need to make these additional adjustments:

- Tone down the white area on the blue bag

- Add shadows under the man's shoes and some rim lighting near the right-sided edges to add more emphasis to him

First, let's work on the bag.

Toning down the white on the bag

I feel that darkening the white on the bag is the best solution:

1. Create a new Layer on the **Man Walking** Layer, and then select the Color Picker to sample the blue color on the bag.

2. Use the Color Picker to select a dark blue color. Use the **Soft Round Brush** to paint over the white space. Reduce the Layer opacity to **82%**, and change the Layer mode to **Multiply** (see *Figure 9.49*):

Figure 9.49 – Tone down the white on the bag with the Brush tool and Multiply

Now that we've toned down the white on the bag, let's add some shadows under the man walking's shoes so that it doesn't look like he's floating above the ground.

Adding shadows under the shoes

The best way to add shadows under the shoes is to use the **Pen** tool to trace the shadow area of the shoes from the original photo of the Man Walking. Let's proceed:

1. First, use the **Pen** tool to trace the shadow of the shoe, make a selection, and fill the shadow with a charcoal black color.

2. Duplicate the shadow Layer, go to **Filter**, add a Gaussian blur, and adjust the **Radius** to **1.4** (see *Figure 9.50*):

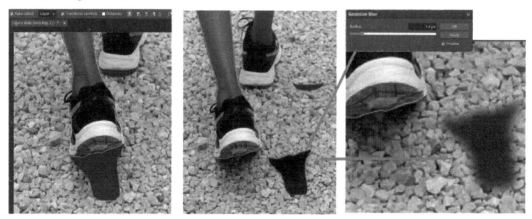

Figure 9.50 – Trace the shoe shadows with the Pen tool and fill the color with black

3. Change the Layer mode to **Pass Through** on the first **Shoe Shadow** Layer. You can see the shadow begin to soften (see *Figure 9.51*):

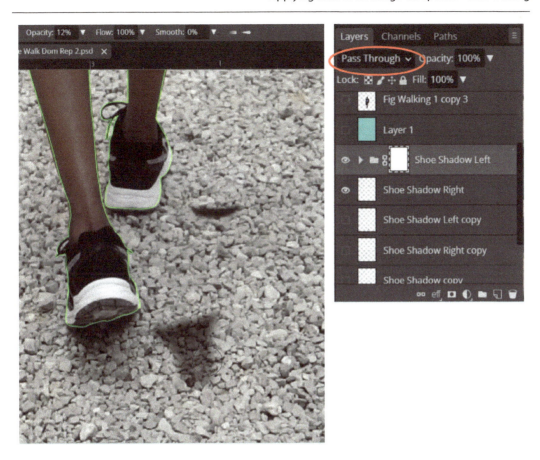

Figure 9.51 – Trace the shoe shadows with the Pen tool and fill the color with black

4. On the second copy of the **Shoe Shadow Left** Layer, change the Layer mode to **Soft Light**, and then change the last **Shoe Shadow Right copy** Layer mode to **Multiply** at **70%** opacity. You can then group the **Shoe Shadow** Layers into a Folder.

5.   Name the Folder **Shoe Shadow**, and then *copy and paste* the *shoe shadows* into the **Image Composite: Man Walking** document (see *Figure 9.52*):

Figure 9.52 – Adjust the shadows using layer modes

> **Important note**
>
> There are a number of ways to create shadows. There will never be one definitive method or approach to adding shadows. Lighting plays a role in how well a shadow appears. Different images may have different lighting and colors, so you may have to study other images to add shadows from your memory and imagination.

Adding the finishing touches

Now that we've added shadows under the shoes, we can start adding the finishing touches to achieve a believable Image Composite in the following steps:

1.   I felt the Trees and Bushes edges were a bit hard, so I will apply a **Gaussian Blur** filter to each one. I will change the **Blur** properties slightly differently for each one as needed. (You can experiment and see how they would look if you blurred them out of focus, simulating a **35 mm** camera shooting with depth of field.)

2.   I duplicated the Rapture, scaled down its size, changed **Opacity** to **70%**, went to the **Edit** menu | **Transform** | **Flip Horizontal,** and changed its position in the sky.

3. I reduced the size of the Man Walking; he needed to be slightly smaller (see *Figure 9.53*):

Figure 9.53 – Duplicate the rapture, resize the man, and apply a Gaussian blur

The last few adjustments gave us a natural and realistic image. Let's make some more adjustments on the Man Walking, to add a little more emphasis to him.

4. Add a bright tan color overlay (**Fill Layer Mask**) to the Man Walking, and paint in some rim lighting along the right-side edges of his ear, shoulder, and arm.

This is achieved by sampling colors from the environment and blending them into the figures.

5. Press *Command* or *Ctrl* + click to make a selection around the Man Walking.

6. Sample the lightest color on the grassy area.

7. Go to **New Layer** | **Fill Layer** to create the mask.

8. Next, switch the Mask Layer mode to **Color Dodge** mode.

9. Next, adjust the Layer Opacity of the color Mask to **21%**.

10. Then, select the **Soft Round Brush**, change **Size** to **60 px**, **Hardness** to **0%**, **Opacity** to **11%**, and paint in the rim light (see *Figure 9.54*):

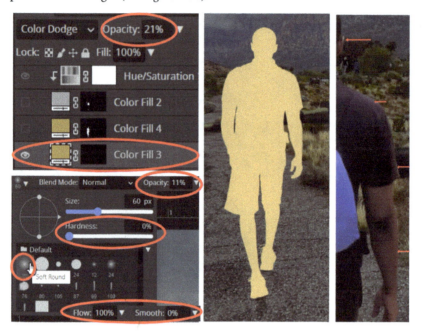

Figure 9.54 – Add a color overlay and paint in the rim light edges

> **Important note**
>
> Change the brush colors back to the default black and white, using the *D* and *X* shortcuts to toggle and paint on the mask.

11. Press *Command* or *Ctrl* + click to make a selection around the Man Walking.

12. Sample the gray color in the clouds.

13. Then, go to **New Layer** | **Fill Layer** to create a Mask.

14. Next, switch the Mask's Layer mode to **Color** mode.

15. Next, adjust the opacity of the gray color Mask's Layer mode to **67%.**

16. Next, select the **Soft Round Brush**, and change the **Size** to **60 px**, **Hardness** to **0%**, and **Opacity** to **100%**.

17. Paint out the gray color except for the bag. This is another way of reducing the color saturation (see *Figure 9.55*).

Figure 9.55 – Add a color overlay and paint in the rim light edges

We've added the **Color Fill Mask** and painted in the rim light edges; now, we can move on to the last steps.

Unify the composite with gradient overlays and adjustment layers

Let's start working on the last steps of the Image Composite of the Man Walking:

1. Place a **Gradient Fill** Layer over the overall image. (It should be placed at the very top of the Layers.)

   Go to **Layer | New Fill Layer | Gradient Fill**.

   I thought about sampling the gray color of the clouds, but I ultimately decided on a blue-gray color to avoid saturated colors and stay consistent with the gloomy look.

2. I edited the gradient colors with the Gradient Editor.

3. Double-click a color on the **Slider** to change it from black to blue, and change the **Opacity** setting to **58%** at the opposite end of the gradient to make it blend from solid to transparent. You can slide the small black dot to adjust how far the transparency can go (see *Figure 9.56*):

Figure 9.56 – Add a Gradient Fill overlay layer over the Image Composite

You can see the slight difference in the Image Composite after we exit the Gradient Editor (see *Figure 9.57*):

Figure 9.57 – The Image Composite with a gradient overlay

> **Important note**
>
> Explore and edit the gradient fill, make duplicates of gradient styles you like, and choose the best look for your project.

I compared what I've done so far and compared it to the other version I did of this when my computer crashed, and I saw that I needed to tweak a few more things.

4. I went back and edited the **Hue** and **Saturation** adjustment Layers on both the **Man Walking** and **LV Mountains Copy 3** layers.

5. First, I *increased* the **Red Saturation** and **Hue** setting on both the **Man Walking** and **Vegas Background** layers. This helps unify them with the slightly warm red color in the skin tone and the grass area.

6. Next, I *disabled* the **Curves 3** adjustment Layer Clipping Mask on the **LV Mountains copy 3** Layer.

7. I *increased* the **Hue** and **Saturation** settings of the **Blue Bag** to give it a deeper, darker blue, which matches the intensity of **Man Walking** (see *Figure 9.58*):

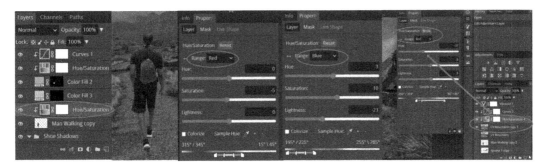

Figure 9.58 – Increasing the Hue and Saturation adjustment layers

8. Next, export the document as a flattened Jpeg file.

9. Open the Jpeg file and select **Save as a PSD**, renaming it *Fig 9.2 Man Walking 9*.

10. Rename the Layer **Man Walking Flat**.

11. Duplicate the Layer and apply a **Grain** filter.

12. Go to **Filter** | **Noise** | **Add Noise** and adjust **Amount** to **10%**. Adding the **Noise** filter over the entire Image Composite gives it both a more unified and gloomy (about to rain) look.

13. I added a **Curves** Adjustment Layer to level the contrast throughout the entire image (see *Figure 9.59*):

Figure 9.59 – Save the image as a new document, flatten the layers, and add a Noise filter

14. Lastly, you can see some differences between the two versions from the changes I made – *increasing* the **Red** hue and saturation in both Layers and *disabling* the **Curves 3** Layer of **Las Vegas background** (see *Figure 9.60*):

Version 1: Before Changes   Version 2: After Changes

Figure 9.60 — Compare version 1 before the changes with version 2 after the changes

I feel that the colors and contrast didn't match up well with the background. The man walking in version 2 has a more unified look with the background after I slightly increased the **Red** hue and saturation, increased the **Blue** hue and saturation on the bag, giving it a deeper, darker blue, and added the **Curves** adjustment Layer to balance the overall image.

Summary

We covered a lot of information in this chapter, *Exploring Advanced Image Compositing Techniques*.

Using two very different Image Composites gave us an in-depth understanding of different ways to approach them. The first one was *Lunch on Table*, where we merged and arranged different photos of dishes, silverware, and tablecloths with simple-shaped objects. This made it easier to dive into making selections, retouching, and swapping backgrounds.

The second Image composite, *Man Walking*, demonstrated how to merge more complex objects – the Iguana, Desert Tree, Bushes, Raptures, and a Man Walking in mountain scenes. We learned how to match colors and moods from images that were taken in different regions, lighting, and weather, using color overlays and Clipping Masks to isolate individual objects, which required individual adjustment Layers. We also simulated shadows, changing Layer modes.

Now that we've covered a large range of skills to create Image Composites, let's move on to *Chapter 10, Text Fundamentals and Styling in Photopea*.

Text Fundamentals and Styling in Photopea

In this chapter, we will explore the fundamentals of text, fonts, and typography. In addition, we will learn how to use the **Type Tool** for artistic text, warped and shaped text, and body text for paragraphs.

Body text often uses character styles for adjusting text size, color, shape, and so on. Body text (paragraph) styles involve things such as kerning, paragraph spacing, margins, applying effects to text, and so on.

By the end of this chapter, you will be able to differentiate between serif and sans serif fonts and will have gained an understanding of historical key points concerning fonts and how fonts have evolved over the years. In addition, you will be able to select fonts and style them to reflect or complement a business, brand, and mood for a logo, design, and page layout.

In this chapter, we will cover the following topics:

- Introduction to typography
- Working with the Type Tool and text styles
- Creating artistic, shaped, and warped text, and more

Introduction to typography

Before we get into exploring text fundamentals and styling in Photopea, I felt we should cover some history first, along with the fundamentals of type, type styles, and fonts. This will help in making choices for different types of image editing projects that may require body text for a book cover, postcard, stationery, decorated apparel, interior book text, web design for your home page and menu pages, and so on.

This section will give you some insight into things to keep in mind and help you make informed decisions while considering the purpose, occasion, mood, and feeling you want your font and text to complement in your images and designs. For instance, you have to consider the font size, thickness, color, and shape, and whether you will be using serif or sans serif as you choose font styles and combinations for your projects. I would encourage you to study typography beyond this section if you want to learn more about it or want to become a graphic designer.

Early beginnings and progression of typography – synopsis

I figured we could explore some of the history of typography before we dive into the tutorials. If you want to learn more about typography, I recommend researching the history of it.

Typography started as images and shapes representing words, sounds, and ideas. Some of the earliest markings of communication are cave paintings that date back as far as 10,000 years ago by early humans (prehistoric era).

The next stages of typography began with the Sumerian writings around 3000 BCE, using styluses and clay tablets. This was followed by Egyptians creating hieroglyphics on stone and, later, using brushes made from plant stems to paint on papyrus paper around 2000 BCE.

Another unique form of writing emerged with Chinese calligraphy at around 1800 BCE. They used symbols, abstracted pictographs, and logograms (symbols representing words – modern-day examples include @, %, and #).

During the Industrial Revolution of the early 1800s, the progression of typefaces and printing soared to new heights. Industrialization, automation, and manufacturing played big roles.

Automated papermaking made paper more affordable and thus ignited a boom for advertising to promote products, and created a need for bolder, larger, decorative fonts such as fat faces and Egyptian faces.

The invention of the penny press by Ben Day made it possible to start an independent newspaper with inexpensive production costs and gain more advertising slots for products in papers. By then, type casting technologies by David Bruce enabled setting type with a type casting machine, 10 times faster than someone could do by hand.

Type and print also played a role during the Arts and Crafts movement – from decorative fonts complementing the art style movements to minimized modern art posters to modern advertising posters and poster ads used for propaganda.

In the late 1960s to 70s, the Late Modern movement was formed, with leaps in possibilities in typography, design, and technology. A form of typography and imagery evolved in urban cultures, where graffiti artists and tag crews began creating stylized typography and images in modern forms of the prehistoric and ancient cultures of Egyptian hieroglyphics, Sumerian writing, and Chinese calligraphy, to name a few.

Now that we have gone over the history of typography, let's dive into the next section on learning about the differences between serif and san serif fonts.

Serif versus sans serif fonts

With the evolution of typography, computers, and technology, we now have thousands of typefaces to choose from. The first step is making sure we can distinguish between the two main font styles (typefaces): serif and sans serif. After that, we can better decide on which to use for each project and why.

Key differences between serif and sans serif

The key differences between the two typefaces are the stems. A serif is a decorative line or mark (also known as a *foot* or *tail*) placed at the beginning and end of each letter's stem. The added marks result in the serif type taking up more space on the page or area of composition.

Sans serif fonts are clean, simply shaped, and consistent, with the same width and taking up less space.

Some of the more popular serif fonts used are Times New Roman, Georgia, and Baskerville. The most used sans serif fonts are Futura, Arial, and Helvetica. There are some other popular fonts from both typefaces I didn't include that you'll discover on your own.

Just as certain colors can trigger a mood or complement your brand, a font can similarly do those things, especially decorative fonts.

Here is an example using serif on the left and sans serif on the right. I wanted to find a font and type to complement the mood of the phrase *Relaxing in my comfort zone.*

Let's start with the example on the left; we know that the colors red and orange trigger the moods of excitement, energy, and aggression. The serif Times New Roman font has sharp edges, and most of the letters are touching each other due to tracking. This gives a sense of tension and stress. That doesn't work for the phrase.

However, the sans serif example on the right has nice tracking (spacing between the letters) so there is no sense of tension. The text is blue, giving a sense of calm, comfort, clarity, and trust. This aligns with the message *Relaxing in my comfort zone.* See *Figure 10.1*:

Figure 10.1 – Using serif (left) and sans serif (right) fonts to complement the subject matter

Let's continue learning about serif and sans serif fonts in the next section.

Usage of serif fonts

Traditionally, serif fonts were the way to go if you wanted your designs, logo, or brand to give a traditional, serious, classical, established, or sophisticated look. Serif typefaces evolved around the 18th century, and have been the go-to for most books, magazines, and newspapers since they are more legible to read with body text. Serif fonts are still used to this day by traditional businesses such as law firms and insurance companies, as well as corporate copy and materials for marketing brochures and **requests for proposal** (**RFPs**).

Usage of sans serif fonts

Sans serif fonts were usually the way to go If you were looking for a modern, non-traditional feel. Things have changed since then and companies such as Google use sans serif fonts to represent their brand.

Sans serif fonts have a simplistic, minimal style, which can give your designs and brand a clean, youthful, and more approachable touch for your viewers.

Companies such as Google are great examples of using sans serif versus the traditional serif look. That goes to show how typeface choices can also be influenced by design trends.

Now that we've covered reasons why you should choose one over the other, let's take a look at the anatomy of a sans serif and a serif font.

Anatomy of serif versus sans serif fonts

The following figure shows an example of a serif font (at the top), and a sans serif font (at the bottom). You can see the serif (extended mark) on the baseline of the A versus the more modern A in the sans serif example. Pay attention to the height of the capital letter, the baseline, where all the type rests, and the descender line just below the baseline; the tails of the lowercase letters such as p, g, y, and q rest on the descender line. Also notice the x-line height for lowercase letters, as well as the crossbar that joins the shape of the capital B, lowercase e, and so on. See *Figure 10.2*:

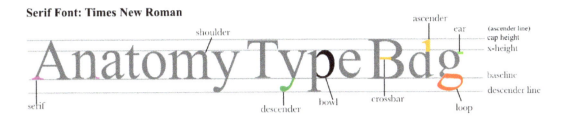

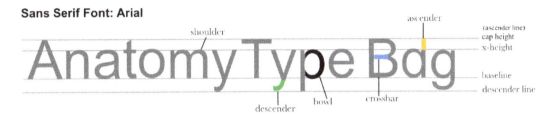

Figure 10.2 – Anatomy of serif and sans serif fonts

Now that we've briefly covered the anatomy of serif versus sans serif fonts, let's move on to the next section.

Working with the type tool and text styles

Photopea's type tool and text styles add another level of creativity to standalone images. You can use text to help tell or sell a story, give context to your images, promote a product, design movie posters, give step-by-step instructions, write a book, design a t-shirt, and so on. Let's explore the type tool and learn how to use it for different project scenarios.

The type tool

Before we dive into the type tool, let's set up a new document:

1. Create a new document with the **Letter** size (**8.5" x 11"**) with **300 DPI** or Dots Per Inch.

2. Select **Type Tool** from the toolbox, or press *T* on your keyboard to activate it. When you click **Type Tool** on the document, you can type out a word horizontally in default mode.

3. A new **Text** Layer with a capital **T** will automatically appear in the **Layers** panel.

4. You will also see the **Type Tool** properties bar displaying the current font, font color, text size, and text alignment appearing at the top near the main menu. See *Figure 10.3*:

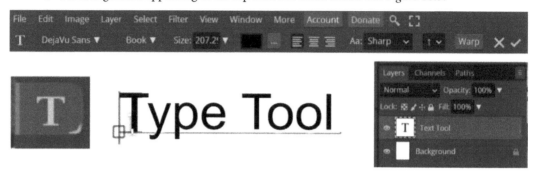

Figure 10.3 – Selecting Type Tool

5. To type out text vertically, click the **Type Tool** and select the **Vertical Type Tool**. See *Figure 10.4*:

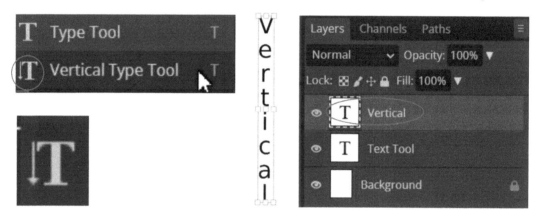

Figure 10.4 – Vertical Type Tool

There are three types of text you can achieve with the **Type Tool**:

*   **Point text** (as covered in *Figure 10.3*) – You can click the **Type Tool** on any area of the canvas to start typing. This is best for short phrases, titles, and headings.

*   **Paragraph text** – This is good for typing out paragraphs (large bodies of text). This is achieved when you click and drag the **Type Tool** on the canvas to create a rectangle (bounding box), then release the box and begin typing. See *Figure 10.5*:

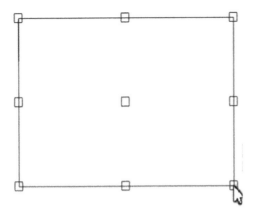

Figure 10.5 – Clicking and dragging the Type Tool to make a bounding box

- You can also add paragraph text inside a shape. Create a shape using the **Pen** tool, or select premade shapes from the **Shape** tool, located in the toolbox.

- Once the shape is made, click on the inside of the shape and begin typing, or copy and paste your paragraph text inside the shape. See *Figure 10.6*:

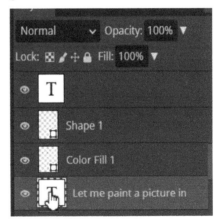

Figure 10.6 – Fitting the text inside a shape

- **Text on a curve** – You can have your text flow along the outside or the inside (opposite) of the curve's path (outline).

    - To do this, create an **Ellipse** shape or circle using the **Shape** tool. Make the shape have no color outline and no color fill. See *Figure 10.7*:

Figure 10.7 – Creating an Ellipse shape

6. Select the layer of the curve first; then, click on the contour of the curve (outline) with the **Type tool** and begin typing on the path of the oval. Two symbols may appear, an **X** and **O** symbol, or you may see a small blue dot. See *Figure 10.8*:

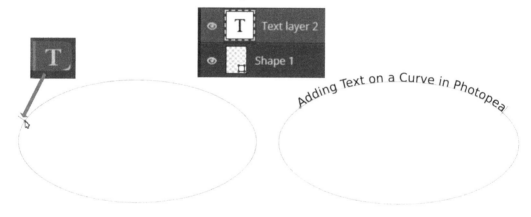

Figure 10.8 – Using the Type Tool and typing on the path of the ellipse

7. Select the **Path Select** tool to drag the blue dot left, right, up, or down to adjust the points, or switch the text to read on the opposite side of the curve. I dragged it around to give you a better idea of the flexibility of moving the text around on the path. See *Figure 10.9*:

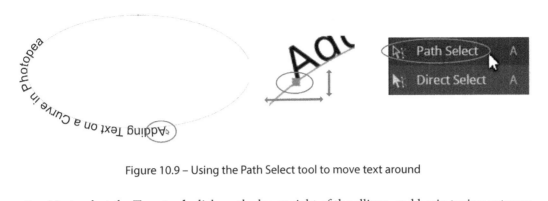

Figure 10.9 – Using the Path Select tool to move text around

8. Next, select the **Type tool**, click on the lower right of the ellipse, and begin typing out your text. See *Figure 10.10*:

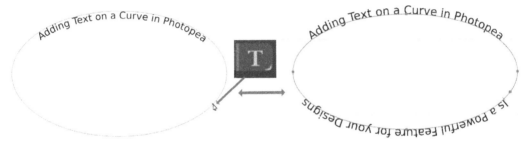

Figure 10.10 – Typing out the text on the lower-right side

9. Once you've finished typing, select the **Path Select** tool and drag the text to the left so it can read from left to right (see *Figure 10.11*):

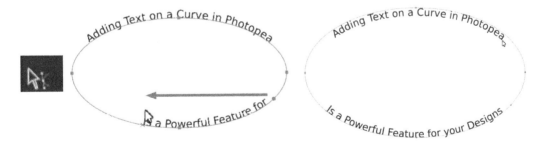

Figure 10.11 – Dragging the text to the left with the Path Select tool

10. Use **Rulers** to create a guide to make sure the text is lined up horizontally on both the left and right (beginning and end of the sentences) for the top and bottom text. See *Figure 10.12*:

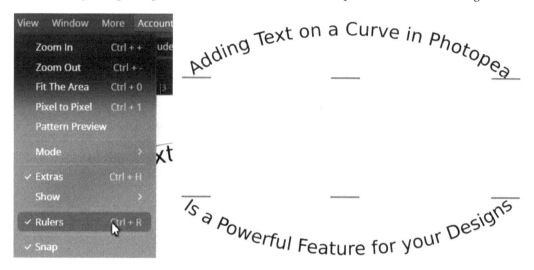

Figure 10.12 – Aligning the text with the horizontal ruler guides

Now that we've covered how to add text to a path and align text with ruler guides, we are ready for the next section.

How to edit a Type layer

Some helpful tips for editing your Type Layers follow.

To activate a new Type Layer, click on the canvas with the **Type Tool**, and the Layer will be created automatically and then locked.

To get out of the Type Layer, you have to confirm or cancel the changes made by pressing the check mark button in the top menu. You can also press the *Esc* key or click the **X** button next to the check mark to abort the changes. See *Figure 10.13*:

Figure 10.13 – Confirming or canceling type edits

You can edit the type color, size, font, alignment, and so on in the properties bar located just below the top menu. See *Figure 10.14*:

Figure 10.14 – Type properties bar

To copy and paste text, you can select the **Type Tool** and drag it over all or part of the text you would like to copy, then press *Ctrl + C* (PC) or *Command + C* (Mac). To paste it into another area of the canvas or inside a paragraph, press *Ctrl* or *Command + V*.

You'll know the text is selected when the black color rectangle highlights a letter or words that you drag the **Type Tool** across. You can also change the text color, size, font, and so on when it's highlighted. See *Figure 10.15*:

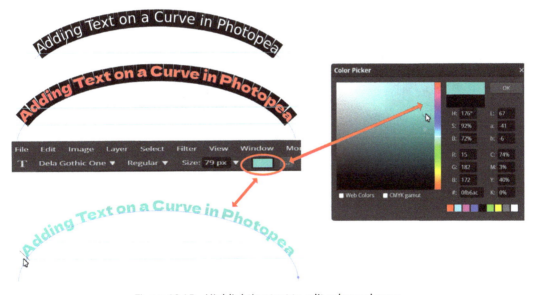

Figure 10.15 – Highlighting text to edit color and more

To move text on a path (curve or other shape) or paragraph text, use the **Move** tool to move the text around.

You can also resize text with the **Move** tool by clicking and holding the mouse button while holding the *Shift* key to resize text in proportion.

> **Important note**
>
> You can also make some edits with text in the **Edit** menu for things such as copying and pasting text, transforming the scale, rotating, skewing, flipping horizontally, and so on. Just make sure the type or text is selected first.

Text styles

Text styles come into play when you need more control over formatting various styles of text. You may need to add paragraph text (body text) to go along with an image or images – for example, a two-page spread for an interior magazine article. You will likely need a large title text, maybe a heading, a subheading, and bodies of text, that may need to be arranged to read in a hierarchy of importance.

Text styles in Photopea have two style parameters you can use to apply to your text. You can access text styles under the **Window** menu and select either **Paragraph** or **Character** to prompt the **Text Styles** panel (when hidden).

It's also located in the sidebar in the example on the right. See *Figure 10.16*:

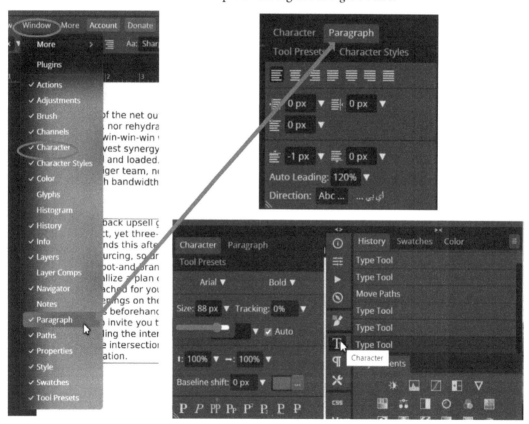

Figure 10.16 – Locating Text Styles from the Window menu or sidebar

The **Character** style gives you the ability to edit the text size, color, font, and so on. The other style parameter is called **Paragraph**. It enables you to align text in a paragraph, spacing between words, and more, which we will go over in more detail next.

Let's look further into the **Character** and **Paragraph** styles.

Character styles

Let's begin our journey on **Character** styles with tracking.

Tracking is a text format that enables you to adjust the spacing evenly throughout a word. You can use tracking to make a word or phrase stand out from the rest of a sentence or paragraph (to give it emphasis) and can be used for designing an ad or logo, and so on. Take a look at tracking being applied to the word **Tracking** in bold letters in *Figure 10.17*:

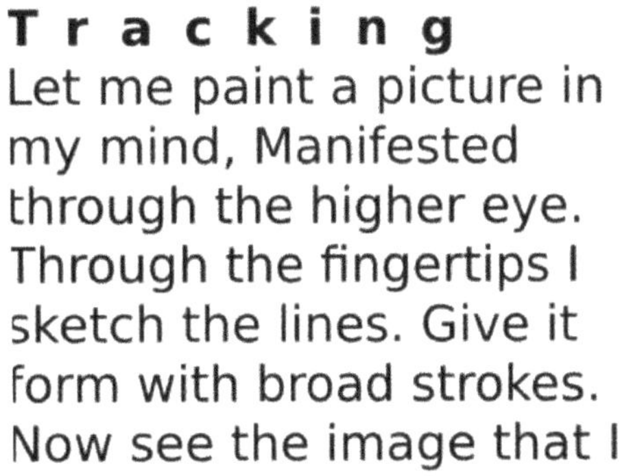

T r a c k i n g
Let me paint a picture in
my mind, Manifested
through the higher eye.
Through the fingertips I
sketch the lines. Give it
form with broad strokes.
Now see the image that I

Figure 10.17 – Using tracking to space out words or bring them closer

> **Important note**
> Try not to change the tracking too much, as it can make your text difficult to read.

Kerning is a common text format used to adjust the spacing between letters in a word. Sometimes, letters created in a font may not space out evenly in a single word or logo and may throw it out of balance or look like a mistake. I purposely changed the kerning on the two red letters of the word **KERNING** to prove a point. Kerning issues will really stand out if it is displayed on a large scale for things such as posters, banners, billboards, and so on. See *Figure 10.18*:

Figure 10.18 – Adjusting kerning to even out the spacing between letters

> **Important note**
>
> Photopea doesn't have a kerning feature, but you can probably fake it by selecting the letters with the **Type Tool** and adjusting them with tracking.

Leading is a text format that allows you to change the spacing between each line of text. You can adjust each line evenly or set up different spacing for each line if you desire (not common with **Paragraph** text, but may work for a specific design). See *Figure 10.19*:

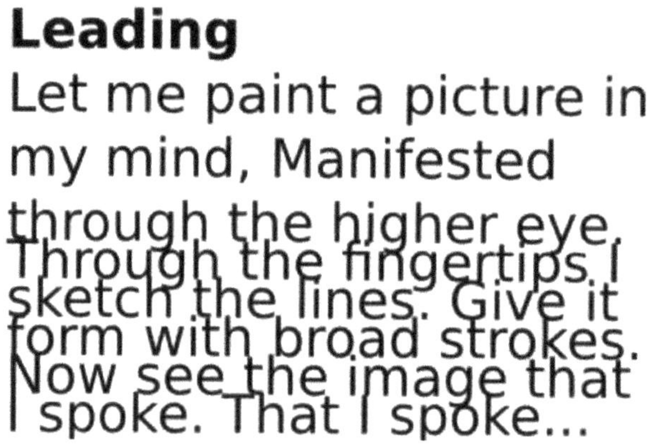

Leading
Let me paint a picture in my mind, Manifested through the higher eye. Through the fingertips I sketch the lines. Give it form with broad strokes. Now see the image that I spoke. That I spoke...

Figure 10.19 – Adjusting space between lines of text with leading

Baseline shift is another text format feature that allows you to move a character or group of characters up or down relative to the baseline in small increments. This feature is great for characters such as fractions (where I show 2 $^{1/5}$ in red color) or when you need to add the TM for trademark, or C for copyright, and so on. See *Figure 10.20*:

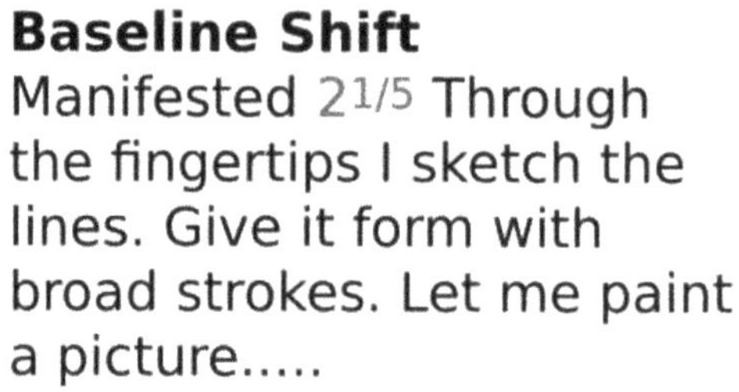

Baseline Shift
Manifested 2$1/5$ Through the fingertips I sketch the lines. Give it form with broad strokes. Let me paint a picture.....

Figure 10.20 – Adjusting the baseline shift for characters such as fractions

Now that we've covered kerning, leading, and baseline shift, we can move on to the next section.

Paragraph styles

At the top, there is the option to align your text left, right, or center. You can also choose the **Paragraph** text to read in a variety of **Justify** options: **Justify Left**, **Justify Right**, **Justify Center**, and so on. **Direction** allows you to switch the direction of reading from right to left (how Arabic is written) versus the default left to right. See *Figure 10.21*:

Creative hunger is running through my veins, blood pumps into the brain, connected to the vertebra. An idea strikes a nerve. Visualizing strong & fascinating images of pixels and energy. Unseen physically, but seen through divine imagery. Meditating, I listen to Him speaking to me. Stimulating the higher eye; Mind, Body, and Soul are working simultaneously, attentively, steadfastly, and intensely. Like a warm illuminating ray of light, its love is gravitating me.This creative energy is flowing through the blood vessels, tendons, and bones. Stimulating

Figure 10.21 – Changing the Paragraph alignment to left, right, and center

Another simple but valuable feature for editing **Paragraph** text is the ability to change the rectangular shape of the paragraph.

To do this, make sure the **Type Tool** is activated, and click in the text once to activate a black bounding box, then drag the corner of the box with the **Type Tool**. You should see a white double-sided arrow to drag it and adjust the shape of the paragraph layout. See *Figure 10.22*:

Creative hunger is running through my veins, blood pumps into the brain, connected to the vertebra. An idea strikes a nerve. Visualizing strong & fascinating images of pixels and energy. Unseen physically, but seen through divine imagery. Meditating, I listen to Him speaking to me. Stimulating the higher eye; Mind, Body, and Soul are working simultaneously, attentively, steadfastly, and intensely. Like a warm illuminating ray of light, its love is gravitating me.This creative energy is flowing through the blood vessels, tendons, and bones. Stimulating .

T Type Tool

Creative hunger is running through my veins, blood pumps into the brain, connected to the vertebra. An idea strikes a nerve. Visualizing strong & fascinating images of pixels and energy. Unseen physically, but seen through divine imagery. Meditating, I listen to Him speaking to me. Stimulating the higher eye; Mind, Body, and Soul are working simultaneously, attentively, steadfastly, and intensely. Like a warm illuminating ray of light, its love is gravitating me.This creative energy is flowing through the blood vessels, tendons, and bones. Stimulating .

Creative hunger is running through my veins, blood pumps into the brain, connected to the vertebra. An idea strikes a nerve. Visualizing strong & fascinating images of pixels and energy. Unseen physically, but seen through divine imagery. Meditating, I listen to Him speaking to me. Stimulating the higher eye; Mind, Body, and Soul are working simultaneously, attentively, steadfastly, and intensely. Like a warm illuminating ray of light, its love is gravitating me.This creative energy is flowing through the blood vessels, tendons, and bones. Stimulating .

Figure 10.22 – Black bounding box to edit the paragraph's shape

If you see a *blue bounding box* around the text, that will either move the text from one location to the other, or it may resize (scale) or rotate it with the **Move** tool. See *Figure 10.23*:

connected to the vertebra. An idea strikes a nerve. Visualizing strong & fascinating images of pixels and energy. Unseen physically, but seen through divine imagery. Meditating, I listen to Him speaking to me. Stimulating the higher eye; Mind, Body, and Soul are working simultaneously, attentively, steadfastly, and intensely. Like a warm illuminating ray of light, its love is gravitating me.This creative energy is flowing through the blood vessels, tendons, and bones. Stimulating .

Creative hunger is running through my veins, blood pumps into the brain, connected to the vertebra. An idea strikes a nerve. Visualizing strong & fascinating images of pixels and energy. Unseen physically, but seen through divine imagery. Meditating, I listen to Him speaking to me. Stimulating the higher eye; Mind, Body, and Soul are working simultaneously, attentively, steadfastly, and intensely. Like a warm illuminating ray of light, its love is gravitating me.This creative energy is flowing through the blood vessels, tendons, and bones.

Creative hunger is running through my veins, blood pumps into the brain, connected to the vertebra. An idea strikes a nerve. Visualizing strong & fascinating images of pixels and energy. Unseen physically, but seen through divine imagery. Meditating, I listen to Him speaking to me. Stimulating the higher eye; Mind, Body, and Soul are working simultaneously, attentively, steadfastly, and intensely. Like a warm illuminating ray of light, its love is gravitating me.This creative energy is flowing through the blood vessels, tendons, and bones. Stimulating

Figure 10.23 – Transforming, rotating, scaling, and moving text with the Move tool

PP, **P**, **P2**, and so on are text style presets that you can use to reformat text quickly. For example, in the Text at the top, the underlined *P* underlines the text. The *P* with the line strike through it will strike through selected text. Lastly, **P2** will change the text you select to a smaller font. See *Figure 10.24*:

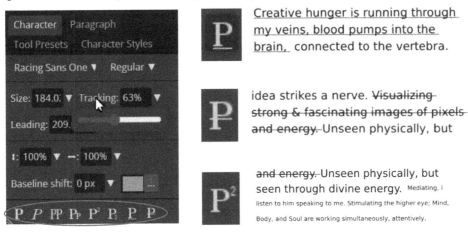

Figure 10.24 – Text style presets

The following figure is an example of a page layout design using text styles, paying homage to a font that I did a couple of years ago. I used a serif font called Nixie-One. The text is based on the history, characteristics, and era of the Dixie-One font. I arranged the text hierarchically, similar to the front page of a newspaper. You wouldn't necessarily use Photopea to lay out text like this. This is a job for a program such as Adobe InDesign or Affinity Publisher. However, you can see how learning about text styles can give your designs a professional look. See *Figure 10.25*:

Figure 10.25 – Designing a page layout using text styles

Now that we've covered text styles and the importance of arranging text in hierarchical order, let's look into how we can expand and keep our designs fresh by expanding your font library. Let's take a look in the next section.

Custom fonts

Expanding your font library with custom fonts can help keep your typography design fresh, add a unique and creative flare to enhance your designs, spark new ideas you can present to future clients, and so on.

Photopea has a huge library of royalty-free fonts and is a great way to begin experimenting and getting acquainted with fonts that can give your designs a fresh and cutting edge. You can also open fonts in Photopea that you may have downloaded from other third-party sites in the format of TTF or OTF files. See *Figure 10.26*:

Figure 10.26 – Using custom fonts to enhance designs

Now that we have a better idea of using custom fonts to expand our font library and enhance our designs, we are ready for the next section.

Websites to download free fonts

There are plenty of online resources where you can download custom fonts for free. Be mindful of legal details on licenses to use each font. Some of the font licenses will allow you to use them for personal projects only, whereas some others may permit you to use them for print, digital, and commercial uses.

Here is a list of some of my favorite sites to download free fonts:

- `www.fonts.google.com/`
- `www.DaFont.com`
- `https://www.1001freefonts.com/`

Now that we've covered custom fonts and where to find free fonts, let's move on to the next section.

Creating artistic, shaped, and warped text, and more

Photopea has some powerful and creative Type tools we can incorporate with our standalone images. Features such as warp text, layer styles, and clipping masks help keep our designs fresh with limitless possibilities. Let's dive in and explore some of the possibilities in this section, starting with warped text and layer styles.

How to warp text and add layer styles

To get started reshaping your text with the **Warp** tool, type out a name or word and choose a font.

Next, select the **Warp** tool located on the right end of the tool property bar and a new tab will open the different **Warp** shape styles for text.

I chose the **Shell Lower** style for the **COOL STYLE** text and adjusted the **Bend** value; feel free to experiment with **Horizontal Distortion** and **Vertical Distortion** and the other **Warp** shapes to get comfortable using it. See *Figure 10.27*:

Figure 10.27 – Using the Warp tool to shape text

Next, select the text with the **Type tool** and then double-click on the text color box to change the color of your choice. See *Figure 10.28*:

Figure 10.28 – Changing the text color

I would recommend you take a look at all the different **Warp** shapes that can be applied to your text, along with experimenting with adding more or less to the bend, horizontal, and vertical distortion that can be achieved with it for your designs.

Now that we've covered the Warp tool, let's dive further into adding layer effects and adjusting text with the **Character** panel in the next section.

Adding layer effects and adjusting text with the Character panel

This section will be similar to the previous section, except we won't warp the text, and this one will require us to make adjustments to the text using the **Character** tab. Follow these steps:

1. To get started, I typed out the words *ARTISTIC TEXT*. The first thing I noticed was the letters were touching (too close together).

2. To fix this, open the **Character** panel and adjust the tracking to space the letters apart. See *Figure 10.29*:

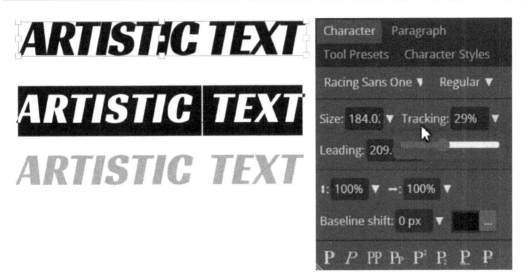

Figure 10.29 – Spacing the letters apart with the Tracking slider

3. After we adjust the tracking, let's create a black stroke around the letters with the **Layer Style** panel.

4. Creating the black Stroke has caused the letters to touch again. See *Figure 10.30*:

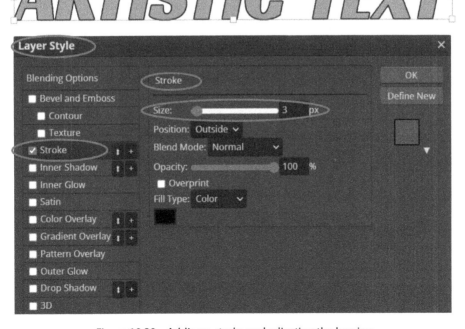

Figure 10.30 – Adding a stroke and adjusting the kerning

5. To fix the issue with the letters touching, turn off the layer styles on the layers and select the **Type Tool** to select (highlight) the letters and adjust the spacing of the letters using the **Tracking** slider. After that, turn on the layer styles for the stroke to see whether the letters are spaced apart evenly. The last two letters to fix were **A** and **E**. Continue adjusting the tracking until satisfied. See *Figure 10.31*:

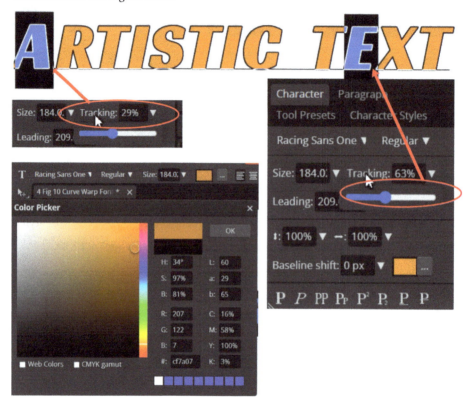

Figure 10.31 – Adjusting the tracking

6. Next, create a new layer below the text and select **Add Fill Color** to fill the layer with a new color. This will help us see more of the layer styles being added to our text. See *Figure 10.32*:

Figure 10.32 – Creating a new layer and filling it with a new color

7. Next, right-click on the **Artistic Text** layer, select **Blending Options**, and add a second stroke in the **Layer Style** panel.

8. Double-click on **Fill Type** to open **Color Picker** and change the color from black to white.

9. Increase the width size of the **Stroke** to **16**. See *Figure 10.33*:

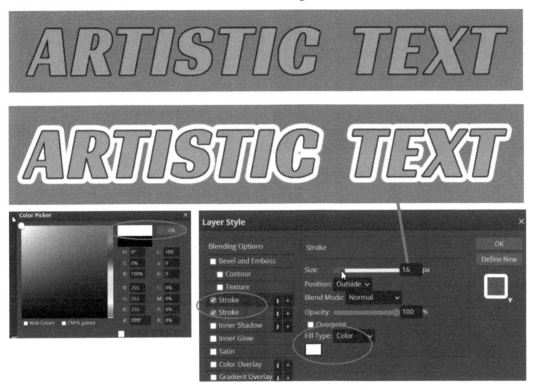

Figure 10.33 – Creating a second stroke and changing the color and size

10. After making the second stroke, select **Drop Shadow**. Double-click on **Drop Shadow** to open the **Drop Shadow** properties to adjust the size, angle, and anything else you want to adjust to your liking (see *Figure 10.34*):

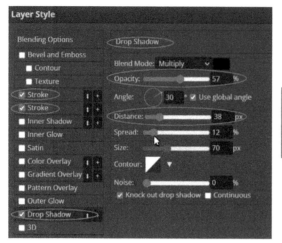

Figure 10.34 – Adding Drop Shadow and making adjustments

11. You can hide and reveal the layer styles that were added to the text by clicking on the **eff** arrow on the right side of the layer to reveal the effects. Click on the small **eff** button to hide or reveal the effect. See *Figure 10.35*:

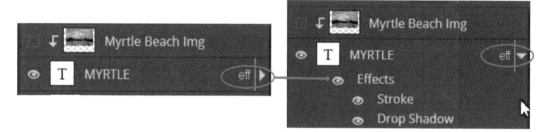

Figure 10.35 – Hiding and revealing Layer Style effects

Continue experimenting with the **Layer Style** effects for styling your text to learn and keep your creative text fresh.

Now we can move to the next section on how to fill text with an image.

How to fill text with an image

Filling text with an image in Photopea is an easy and creative feature that can quickly grab your viewers' attention, and clearly get your message out in a memorable way. Let's create one with a photo I took on vacation at Myrtle Beach:

1. Let's create a horizontal **Letter** size document at 300 DPI. Then, type out *MYRTLE* on one layer and *BEACH* on another. Choose a bold, thick, or heavy font to get as much of the image inside the text as possible. See *Figure 10.36*:

Figure 10.36 – Typing out MYRTLE and BEACH on two separate layers

2. Next, select the **Move** tool to select the word **BEACH**, and hold the *Shift* key while dragging the mouse to scale down the text proportionally to about one-third the size of **MYRTLE**. See *Figure 10.37*:

Figure 10.37 – Scaling down BEACH with the Move tool

3. Next, select the **Type Tool**, click and drag it over the word **BEACH**, and use **Tracking** to space out the letters in **BEACH**. See *Figure 10.38*:

Figure 10.38 – Selecting BEACH with the Type Tool and adjusting Tracking

4.   Next, select the **Type Tool**, click and drag it over the word **MYRTLE**, and use **Tracking** to space out the letters in **MYRTLE**. See *Figure 10.39*:

Figure 10.39 – Selecting MYRTLE with the Type Tool and adjusting Tracking

5.   Next, drag the **Myrtle Beach** photo into the document. The black square means you need to right-click and rasterize the layer. See *Figure 10.40*:

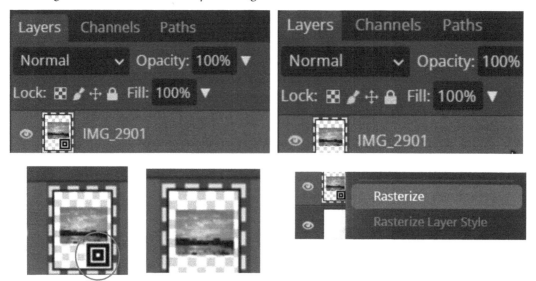

Figure 10.40 – Opening the photo and using Rasterize

6.   Duplicate the photo, then place one photo on top of the word **MYRTLE** and the other on top of the **BEACH** text layer (hide this photo for the time being). Now, right-click over the first photo and add a clipping mask over the word **MYRTLE**. See *Figure 10.41*:

Figure 10.41 – Duplicating the photos and placing them above the words

7. We can see that the photo is placed inside the word **MYRTLE** with the clipping mask. See *Figure 10.42*:

Figure 10.42 – Place the photo inside the word using the clipping mask

8. Use the **Move** tool to drag and position the best part of the image inside of the text. See *Figure 10.43*:

Figure 10.43 – Using the Move tool to position the picture inside of the text

9. Next, right-click over the **MYRTLE** text layer, go to **Blending Options**, and add a black **Stroke** and **Drop Shadow** to make **MYRTLE** more legible to read. See *Figure 10.44*:

Figure 10.44 – Creating a stroke and drop shadows around MYRTLE

10. Next, select the **Line** tool located under the **Rectangle** shape tool in the toolbox, and create an underline between the two words. See *Figure 10.45*:

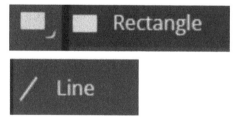

Figure 10.45 – Selecting the Line tool and creating an underline

11. We have created our very own filled-in text with an image. Take a look at the finished design. See *Figure 10.46*:

Figure 10.46 – The finished example for filling in text with an image

This completes this section and the last example for this chapter.

Summary

In this chapter, we delved into some key periods that gave us an understanding of how typography began and evolved. We gained an understanding using the **Text tool** for artistic, warped, and shaped text.

As we ventured further, we gained insight into the power of body text. The **Type tool**, character styles, and paragraph panel provided us with the means to handle formidable paragraphs and extensive bodies of text with ease. We discovered the art of employing character styles to fine-tune the size, color, shape, and other attributes of our text. Additionally, we explored the realm of paragraph styles, where concepts such as kerning, paragraph spacing, and margins offered us unparalleled control over the layout and presentation of our textual creations.

We delved into the intricate nuances of font selection, discovering how a well-chosen typeface can elevate our designs to new heights. With a discerning eye, we learned how to identify fonts that resonate with the message and aesthetic we aim to convey. Through experimentation and practice, we learned how to fill text with an image, breathing life into our words, and enabling us to clearly communicate a message creatively.

In conclusion, our exploration of text, fonts, and typography has equipped us with a heightened understanding of the foundational elements that produce effective visual communication. Armed with this knowledge, we are ready to execute new creative endeavors, confident in our ability to craft visually stunning textual designs.

Let's move on to *Chapter 11, Pre-Designed Templates, Extra Effects, and Features Overview,* where we will explore Photopea's pre-designed templates, extra effects, and features overview.

11

Pre-Designed Templates, Extra Effects, and Features Overview

In this chapter, we will explore the resourceful library of pre-designed templates in Photopea that you can customize and make your own. The templates are perfect for designing social media posts, websites, presentations, and so on. We will create animations to animate logos, create text from scratch, and also learn how to apply animated templates to our logos and content. In addition, we will look at the **Vectorize Bitmap** feature (which I can't believe is free, let alone available, in an image editing software like Photopea). This feature usually comes with vector-based programs such as Illustrator and CorelDRAW.

By the end of this chapter, you'll be able to design social media content with ease, create, open, and export animations, and have the power and flexibility to convert your bitmap images into vector graphics.

For this chapter, we will cover the following topics:

- How to open and edit templates
- Creating, opening, and exporting animations
- Creating an animations in Photopea
- How to vectorize a Bitmap image

How to open and edit templates

The **templates** that are included in Photopea are free to use and are constantly updated with new content by other artists and designers. Photopea gives you 4 categories:

- Mockups

  - Mobil & Web, Prints, Product.

- Social Media

  - YouTube, Instagram. Facebook, Memes, Signs. Animations

- Print

  - Cards, Invitations, Resumes, Diploma/Certificate

- Other

Each category has a sub-category, offering hundreds of designs to help you find a specific template that tailors your needs and message quickly.

You will have instant access to mockups to promote t-shirt designs and other apparel mockups, book covers, a wall sign for your business, and so much more.

Photopea also has custom templates that you can use to set up designs for printed letters, flyers, posters, and business cards. In addition, you have templates for standard computer screens, tablets, and mobile phones

This allows you to get started using the program and making content for your needs very quickly. In the next section, we'll create a design using an Instagram template provided by Photopea and make it our own.

Customizing an Instagram template as your own

Let's get started by customizing a template in this 16-step process:

1. The first thing we need to do is launch Photopea and choose the **Templates** button to the right of the **New Project** and **Open From Computer**.

2. Once you're in the **Templates** window, click on **Instagram** in the **CATEGORIES** section located on the left side.

3. We will select the **Instagram Ad Post** option that says LEARN. MASTER. LEAD., with brown and yellow colors (see *Figure 11.1*):

Figure 11.1 – Select the Instagram Ad Post template

Important note

Note that each template gives you details on who designed or created the template, how long ago, and how many times it's been used by others.

4. Once the Instagram template is open, save it as a PSD file and rename it to your liking.

5. Next, find a photo of a person you would like to replace the lady included in the Template.

    I used an image of a woman in braids. I removed her from the background, using the **Pen** tool to make the selection, and copied some of her braids to give her fuller hair (see *Figure 11.2*):

Figure 11.2 – Finding a photo of someone to replace the lady in the template

6. Once she is in the document as the active layer, go to **Edit | Transform | Flip Horizontally** so that she is facing the text. This will guide the viewer's eye to the message and call to action.

7. Next, type in the text you would like to replace the template text with (see *Figure 11.3*):

Figure 11.3 – Flip the photo of the lady to face the text

8. Now, we can begin to customize the colors, beginning with the red **LEARN MORE** button.

9. In order to change the color of the button and circles, we have to select the button layer and then the **Shape** tool.

10. Once the **Shape** tool is selected, the **Properties** bar will appear at the top underneath the main menu. You will see the current color of the shape that's currently selected.

11. To change the color, *double-click* on it to prompt the **Color Picker** panel (see *Figure 11.4*):

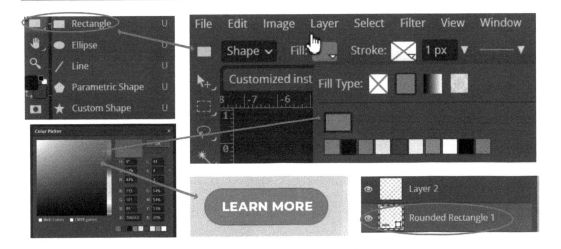

Figure 11.4 – Select the Shape tool and change the LEARN MORE button

I changed the red button to purple for now (see *Figure 11.5*):

Figure 11.5 – Change the button to purple

12. Next, let's change the color of the background square to purple. Instead of selecting the shape to change the color, go to **Edit | Fill | Custom Color** to fill in the color.

13. I also changed the brown circle to light purple (see *Figure 11.6*):

Figure 11.6 – Change the background with Edit | Fill to purple

14. Next, let's break up some of the purple in the background, by filling it in with yellow and filling in the slightly smaller square above the purple to yellow.

15. Then, let's shrink down the light purple circle behind the word **GROW** so that the edges of the circle aren't so close to the edge of the purple square. Designers and artists call these tangents; when elements are too close to the edge of your documents or objects, it can cause a distraction or tension, or lead the viewer's eyes off the page. The goal is to draw the viewer to the page and read the message.

16. The last thing we should do is lighten the large yellow circle and change the purple button to more of a red-violet , making it stand out a little bit more as the call to action.

I think those changes are ideal for an Instagram post (see *Figure 11.7*):

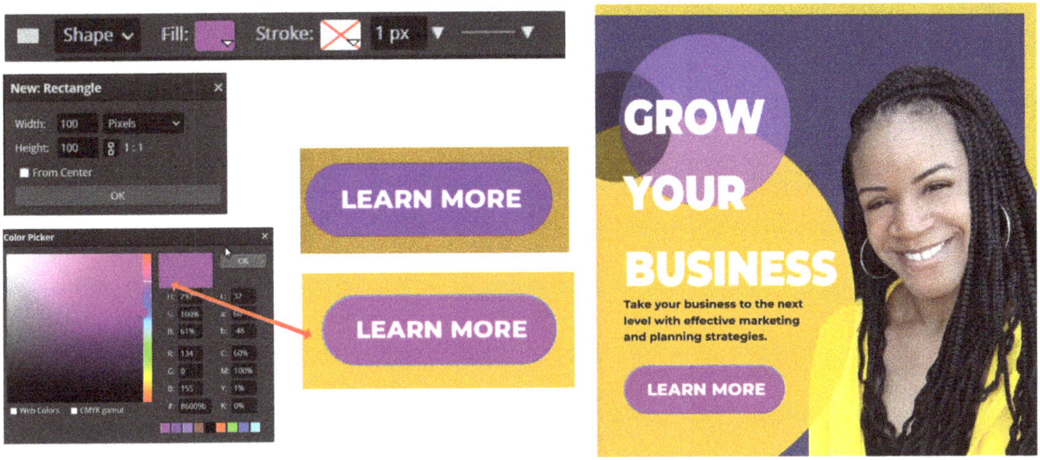

Figure 11.7 – Change the button and circle and add a yellow background layer

Important note

Remember to experiment. The Instagram ad could have gone in an entirely different direction. You can make it more fun with bright colors, use a fun decorative font with a party theme, or have a person with a different facial expression, wearing different clothes for a beach theme, and so on. The options are limitless.

That covers the section on how to customize an Instagram template. Let's move on to the next section.

Creating, opening, and exporting animations

We've covered a great number of features and tutorials thus far in Photopea. For this section, we will cover the basics of how to animate. Being able to animate our still images for social media content and create ads adds another dimension to creating content. It also adds another level of storytelling that may potentially catch a viewer's eye or interest more efficiently when executed well.

Most animated still pictures are created as GIF files. **GIF** is an acronym for **Graphic Interchange Format**. GIFs are still images typically used to display graphics and logos on the web, especially on social media platforms that feature animated memes.

In order to animate GIFs, you need to animate them with a program that supports .gif files and has animation capabilities.

Animation is the ability to simulate movement, or motion, by creating a series of drawings or photos sequentially.

Since the human eye can only capture an image for about 1/10 of a second, multiple images moving at a fast pace create the illusion of a single moving image or frame.

Photopea can open and save animated GIFs, but it can also open and export popular animated files such as .webp files and .apng files. Each file format has its pros and cons.

Creating an animation in Photopea

Creating an animation in Photopea is an easy and straightforward process. Staying organized, naming your layers, and managing the number of files and file sizes are key to keeping your animation workflow smooth, for the most part.

Let's begin our first animation:

1. To begin, create a new document from **Templates** and select the **HD 1280 x 720 px** screen.

2. Next, select and *drag* the **Ellipse** tool and create a small circle.

3. Rename the Layer **_a_,1000 Xsm Circle1** (see *Figure 11.8*):

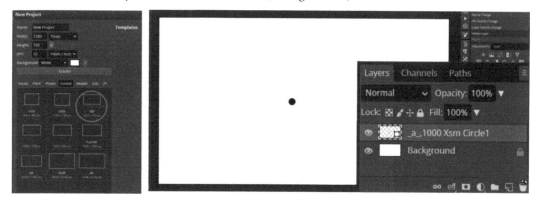

Figure 11.8 – Select the full 1280 x 720 px document template and create a circle

---

Important note

Each Layer has to begin with _a_ , and then a name; otherwise, that Layer won't show up in the animation. The 1,000 part stands for a 1-second frame that we will use as a filler for now.

---

4. Next, *duplicate* the **_a_,1000 Xsm Circle1** Layer, *rename* it **_a_,1000 Sm Circle2**, and *slightly enlarge* the ellipse.

5. To enlarge it, select the **Move** tool, and *drag* the *corner* of the ellipse while *holding* the *Shift* and *Alt* keys so that the ellipse stays centered. You can lighten the transparency to check the size and proportions of both circles if unsure.

6. *Duplicate* the **_a_,1000 Sm Circle2** Layer, rename it **_a_,1000 Md Circle3**, and *slightly enlarge* the ellipse.

7. Next, *duplicate* the **_a_,1000 Md Circle3** Layer, rename it **_a_,1000 Lg Circle4**, and *slightly enlarge* the ellipse.

8. Next, duplicate the **_a_,1000 Lg Circle4** Layer, rename it **_a_,1000 XL Circle5**, and *slightly enlarge* the ellipse (see *Figure 11.9*):

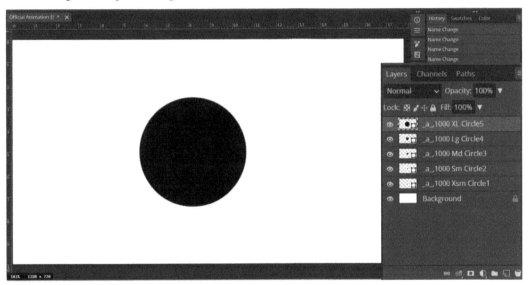

Figure 11.9 – Duplicate _a_,1000 Xsm Circle1 and enlarge and rename Duplicate _a_,1000 Sm Circle2

Important note

Remember that the order the animation plays starts with the first Layer, which begins from the bottom Layer going up.

Now that we've set up all the Layers with the five different-sized circles, we can add or adjust a time length for each Layer.

9. To set up the timing for each Layer, *rename* the Layers as follows.

Make sure to *add* a comma and 1000 of greater or lesser for the time frame after _a_,:

- **_a_,2000 XL Circle5**

- **_a_,1000 Lg Circle4**

- **_a_,1000 Md Circle3**

- **_a_,1000 Sm Circle2**

- **_a_,1000 Xsm Circle1**

See *Figure 11.10*:

Figure 11.10 – Edit the layer names with the time length for each one

---

Important note

1000 stands for *1 second*, 2000 stands for *2 seconds*, and so on. If I need a frame to last less than a second (for example, half a second), I can change the number from 1000 to 500, and type it as **_a_,500 Circle1**.

Now, we are ready to test the animation.

10. *Select* the very first Layer, **_a_,1000 Xsm Circle1**, because that is the order in which the animation will read, export, and playback the frames or steps of the animation

11. Go to **File | Export as GIF** to see the preview of the animation before we export it as an animated GIF.

12. Set the animation speed to **437%**. You can adjust the speed to what looks best for different projects.

13. After testing the animation out, I felt I needed to duplicate some of the Layers to help fill in the gaps (transitions) between the Layers to make the circle enlarge smoothly.

Let's duplicate the following **Layers**:

- **_a_,2000 XL Circle5**

- **_a_,500 Lg Circle4 copy**

- **_a_,1000 Lg Circle4**

- **_a_,1000 Md Circle3 copy**

- _a_,500 Md Circle3

- _a_,1000 Sm Circle2 copy

- _a_,1000 Sm Circle2

- _a_,1000 Xsm Circle1

See *Figure 11.11*:

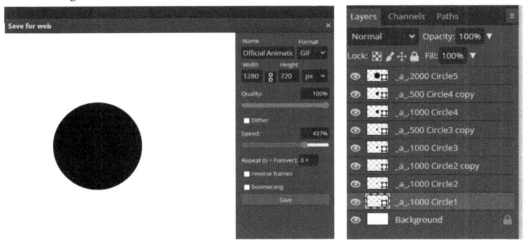

Figure 11.11 – Duplicate the layers to increase smoothness in the animation

Testing the animation, duplicating the Layers, and increasing the speed helped smoothen the transitions or steps between the circles.

> **Important note**
> This animation tutorial applies to animating text, images, characters, pictures, drawings, logos, photos, simulated rain and snow, and so on.

14. Now, we can open the GIF file we exported to see whether it can reopen as separate Layers.

It can open up as separate Layers, but they are renamed, and the first Layer is the only Layer visible; the rest are hidden. You have to click the Layer box to reveal them again (see *Figure 11.12*):

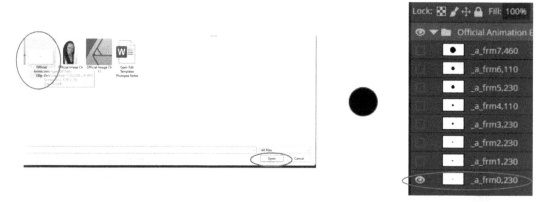

Figure 11.12 – Open the exported GIF file to view the newly renamed layers

That covers the basics of how to create an animation in Photopea. In the next section, we'll look at one more approach to creating an animation in Photopea – by customizing a premade template.

Opening and customizing an animation template in Photopea

We can create an animation quickly in Photopea using one of the free animated templates:

1. To get started, launch Photopea, *select* **Templates**, and *click* **Animations** under **CATEGORIES**.

2. Click on the template that says **logo animation** to open it (see *Figure 11.13*):

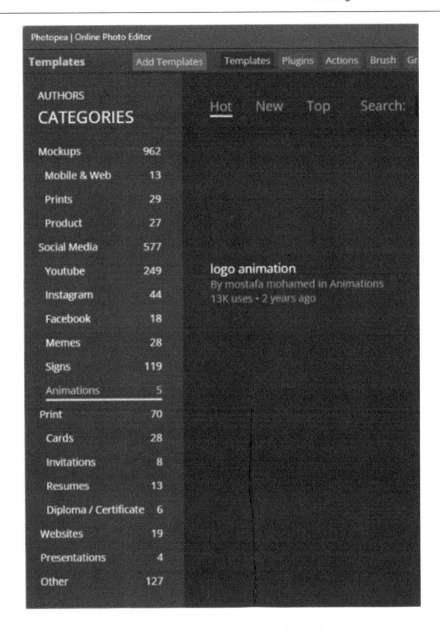

Figure 11.13 – Open the exported GIF file to view the newly renamed layers

3. Once it's open, a PSD file with instructions on how to use the template will appear (see *Figure 11.14*):

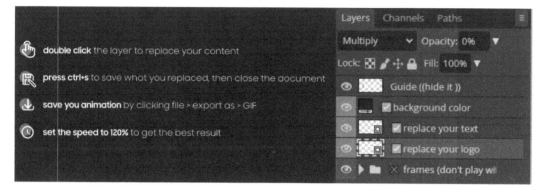

Figure 11.14 – The animation template opened with instructions on how to customize it

4. Go to the **File** menu, select **Save as PSD**, and *rename* the file with a name you prefer for the project. Next, follow these instructions.

5. *Double-click* on the background color if you want to change the color. I will change it to black (see *Figure 11.15*):

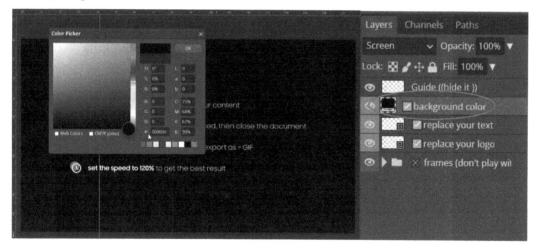

Figure 11.15 – Double-click on the background color layer to change the color

6. Next, *double-click* on **replace your text** and replace it with your own. I exported my text as a PNG file and dragged it into the program.

7. I placed the text above the template text and converted the text into a Smart Object. Go to the **File | Save (Smart Object)** or press the *Ctrl + C* shortcut (see *Figure 11.16*):

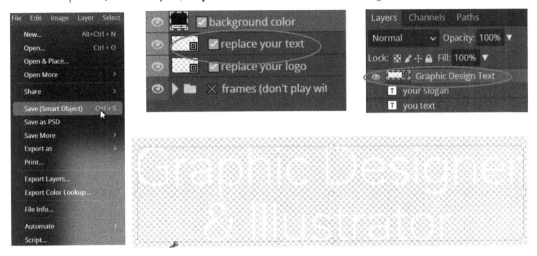

Figure 11.16 – Double-click on replace your text to customize and save the logo layer as a Smart Object

8. Now, you can double-click on the replace your logo Layer and add your own logo.

   Save the logo Layer as a Smart object (see *Figure 11.17*):

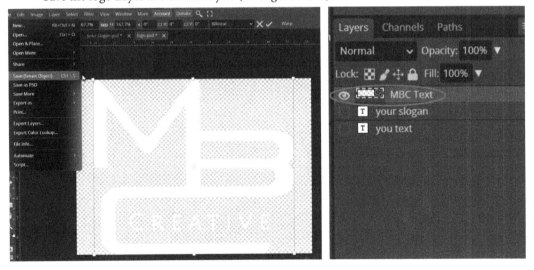

Figure 11.17 – Save the logo as a Smart Object

9. Next, I recommend hiding the Layer named **Guide ((hide it))** as seen in the example *Figure 11.15*; so that the template logo doesn't show up in your logo.

10. Now, we are ready to save the animation; go to the **File** menu | **Export as** | **GIF**, and set the speed to **120%**:

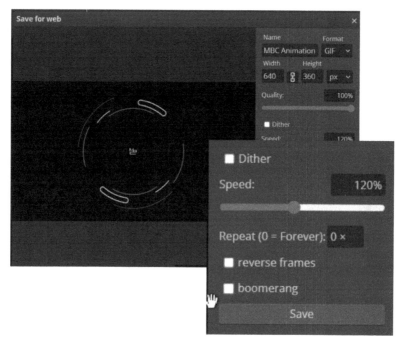

Figure 11.18 – Export the logo as a GIF and adjust the speed

Here are a few shots of the animation preview in GIF export mode.

Figure 11.19 – Screenshots of the logo animation

That covers this section on the animation process; let's move on to the next section.

How to vectorize a Bitmap image

Vectorizing a Bitmap feature is an amazing feature that I mentioned at the beginning of the chapter and is one that usually comes with vector-based programs, such as Illustrator and CorelDRAW. To be able to trace an image such as a logo, line drawing, or one with flat colors depends on how well the image displays can be traced quickly in Photopea. You can save hours of time trying to retrace an image with the **Pen** tool.

Remember that we covered the basics of vector-based images in *Chapter 1*, *Taking Your Design and Editing to the Next Level with Photopea*. We know that vector images and graphics can be infinitely scaled up or down, they are usually smaller file sizes than hi-res raster files and are best for spot colors, logos, technical drawings, 3D and 2D animation, specialty printing, embroidery, color separations, and CAD drawing.

Let's take a look at the avatar that I created as a vector-based artwork, with light raster brush strokes:

1. I exported it as a JPG and will open it in Photopea as a raster image. Note that it shows up as a flat image or one Layer.

2. Go to **Image** on the top menu and *select* **Vectorize Bitmap...** (see *Figure 11.20*):

Figure 11.20 – Open the image and go to Image | Vectorize Bitmap...

3. The **Vectorize Bitmap** window will open with the original image on the left, and the traced vector art will appear on the right. The vectorized trace calculated 17 colors, and the vector art looks rougher than the original on the left (see *Figure 11.21*):

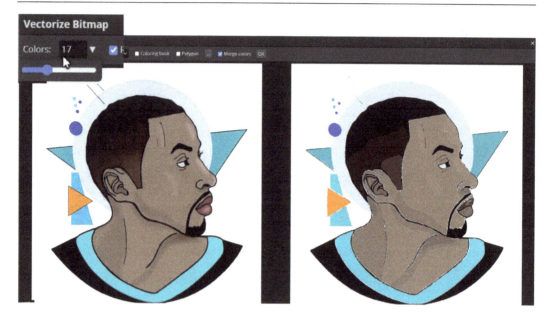

Figure 11.21 – The original image on the left and the vectorized art on the right

4. We can try to smooth out the vectorized image by *sliding* the **Colors** bar to the left to reduce some of the colors. I reduced it to **5** colors. This will combine similar colors together, resulting in smoother lines and shapes (see *Figure 11.22*):

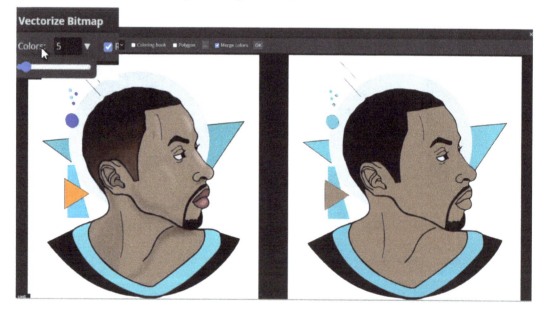

Figure 11.22 – Smooth out the vectorized traces with the slider

5.   Some things that changed after we vectorized the original image are as follows:

  • The original image Layer is no longer available

  • All of the shapes converted to flat colors are now independent vector Layers

  • Now, you can show and hide each vector shape (see *Figure 11.23*):

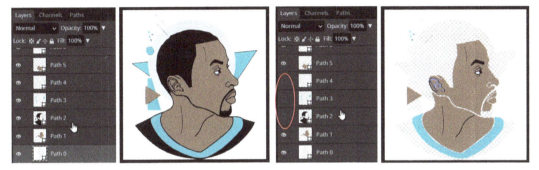

Figure 11.23 – The vectorized shapes are converted to independent vector layers

6.   If you need to export the vector art into other vector-based programs such as Adobe Illustrator, CorelDRAW, Affinity Photo, or Inkscape, you can export the file as a PDF or SVG file with the Vector data intact (see *Figure 11.24*):

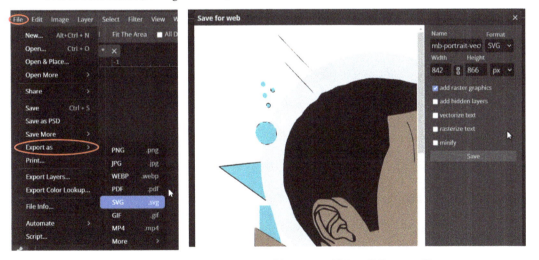

Figure 11.24 – Export your vectorized bitmap as PDF or SVG vector files

That covers this section on how to vectorize bitmap images. Now, it's time to sum everything up. We covered a great deal of information in this chapter.

Summary

In this chapter, we covered some really helpful information, tutorials, and resources that can help jump-start our design flow very quickly with Photopea.

This consisted of pre-designed ready-to-use templates built right into Photopea. Photopea gives you 19 categories and hundreds of design templates so that you don't have to browse and search online all the time. The templates library includes mockups, templates that you can customize for Instagram, YouTube, and Facebook, memes, and templates that you can use to set up designs for printed letters, flyers, posters, and business cards.

From these templates, we customized an Instagram ad template by swapping out text, colors, and a photo for our own. From there, we created a simple animation using the **Ellipse** tool.

Following that, we customized an animation template that gave us instructions on how to use it correctly, and lastly, we learned how to open an image such as a simple drawing or logo that we may need to convert into vector artwork using the **Vectorize Bitmap** tool.

Now that we've covered all the sections of this chapter, let's move on to *Chapter 12, Advanced Color Techniques*.

Part 3:
Drawing Figures, Creating a Logo, and Other Features

This part will explore the library of pre-designed templates that can be customized, as well as advanced color techniques. We will add color to a black-and-white image and explore channels. We will also cover beginner techniques on how to draw, as well as advanced techniques for drawing a character. From there, we will learn how to create a logo for your brand, followed by understanding layer comps, using smart layers, and creating a portfolio to advance your career.

This part comprises the following chapters:

- *Chapter 12, Advanced Color Techniques*
- *Chapter 13, Bonus: How to Draw and Paint a Figure and Character*
- *Chapter 14, Bonus: How to Create a Logo*
- *Chapter 15, Tips, Tricks, and Best Practices*

12
Advanced Color Techniques

In this chapter, we will dive into part two on color in Photopea, using advanced color techniques. We will learn how to add color to a black and white image using solid color fill adjustment layers in conjunction with masks, which can be a great technique for colorizing black and white artwork and old black and white photos while maintaining the true textures of clothing, skin, and other surfaces found on objects in the photo.

In addition, you will learn how to properly use a color photo as a reference guide. This will help you learn how to sample colors with the eyedropper tool in conjunction with using the color picker, and the experience may extend your color vocabulary and help you develop your own color palettes and swatches.

We will also learn how to replace a specific color in your image with the ease of simply adjusting a few color properties, a process that would take much longer if we had to create masks to paint the color and make adjustments.

Another feature we will cover is how to manually select and change colors effortlessly, without losing the original texture of the object or element that needs the color change. This can be useful for quickly showing a variety of looks to a client for a presentation, or deciding on what color schemes work best for a product.

Lastly, we will learn how to modify channels, which will give you a better understanding of what channels are and what they are used for, and will build up your confidence to use them for advanced purposes involving printing and color separations.

In this chapter, we will cover the following topics:

- Adding color to a black and white image

- Replacing a particular color

- Using the Color Replacement brush tool

- Modifying channels

Adding color to a black and white image

Have you ever wanted to add color to a specific area of your image? Or bring some old black and white photos to life with a splash of color? You may have some old family portraits of great-grandparents or even great-great-grandparents, for example. There's something about color that can add another level of storytelling and connection to what we see and experience in life.

Photopea has an arsenal of photo editing tools to bring your black and white images to life. We will explore some easy methods to colorize your black and white photos using vector masks as solid color fills. This will enhance your editing skills and train your eyes to balance contrast and color saturation. You'll also master matching colors using color photos as references, changing color schemes, and applying other image enhancements involved in colorizing a black and white image.

Now that we have a better understanding of what we'll be covering in terms of advanced color techniques, let's get started!

Find a black and white image or convert a color photo to black and white

Before we get started, you need to find a black and white image you want to work with. If you don't have one that you like, you can use my example.

Another option is to convert a color photo of yours into black and white in Photopea. Here are the steps for converting a color photo to black and white:

1. Drag the color photo into Photopea and create a duplicate **Color Photo** layer as a backup.

2. Next, go to the **Image** menu, select **Adjustments | Black & White...**, and then press **OK**. Rename the new black and white layer **BW Image**. See *Figure 12.1*:

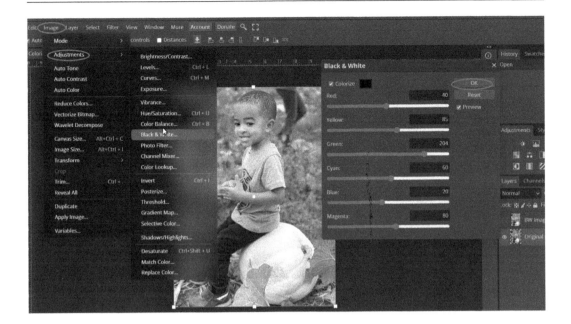

Figure 12.1 – Convert your color photo into a black and white image

Now that we've converted our image to black and white, let's learn how to colorize skin tones.

Colorizing the skin tone

> **Important note**
>
> Don't focus too much on trying to replicate this exactly how I did it. Also, depending on your skill level, this may or may not seem like a lot of steps, but it's better for you to practice with the steps and learn by repetition.
>
> Lastly, keep in mind that a different photo used would not need half of these steps because the image may be simpler. For example, just using a headshot of a boy would mean that you wouldn't have to worry about colorizing the shoes, clothes, or background to the degree required here.

To colorize the skin tone, follow these steps:

1. Select **New Adjustment Layer | Color Fill** to add a new layer above the **BW Image** layer used in *Figure 12.1*.

2. Use the eyedropper tool to Sample an area of the skin color from the original color photo.

3. Change the layer mode to **Color**, and lower the layer **Opacity** to around **65** or **70%**.

4. The image of the boy will have an orange tint from lowering the **Opacity**. See *Figure 12.2*:

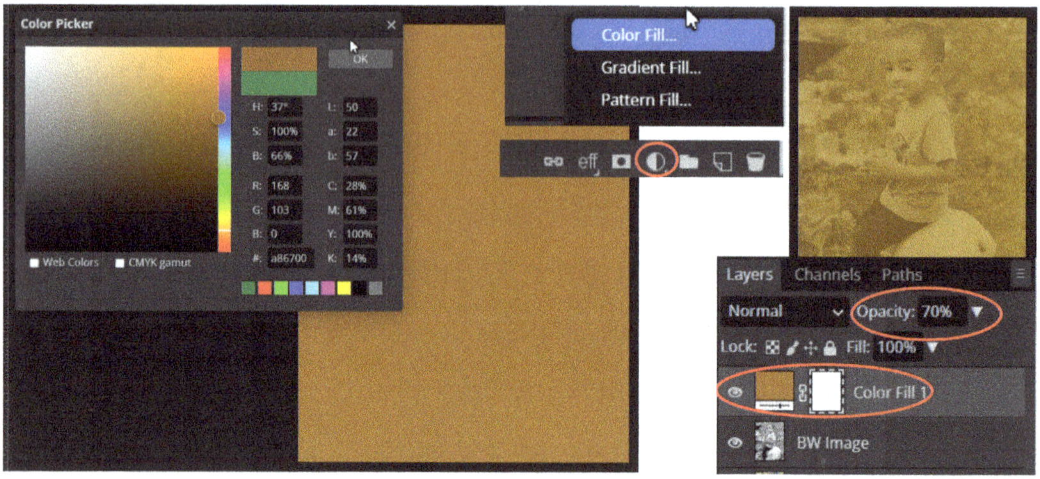

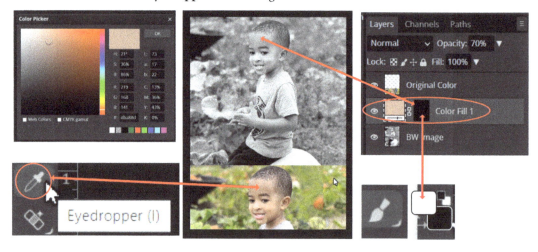

Figure 12.2 – Create New Adjustment Layer | Color Fill

5. Next, click the white square on the **Color Fill 1** mask Layer to activate mask mode.

6. Press *Ctrl + L* on the **Color Fill 1** mask Layer to invert the *white mask* to a *black mask*. The orange tint will change to a black and white image.

7. Now, press *B* to activate the **Brush** tool, and change the foreground color to *white* so we can paint the tan color over the black and white skin.

8. Before we begin painting, open the color photo reference we are working from to sample the skin color with the **Eyedropper** tool. See *Figure 12.3*:

Figure 12.3 – Invert the Color Fill 1 mask to black, and set the Brush tool to a white foreground color

9. Next, rename the **Color Fill 1** Layer to **Skin Tone**. This will help organize and identify each layer as we continue to work on each color.

10. *Double-click* on the image layer **Skin Tone** Layer to open the **Color Picker** tab and use the *color picker* to sample the subject's skin tone. Try to match that color in the **Color Picker** window.

11. Next, click on the black square of the **Skin Tone** mask Layer to activate it for painting.

12. Change the Layer mode to **Color** and begin painting in the skin tone.

13. Start out with a small-sized brush to paint along the edges of the face and neck, and around the eyes and lips. You can paint over the eyebrows and lashes; avoiding the hair and clothes.

14. Afterward, you can increase the size of the brush to paint in the larger areas of the face. Repeat this process for the arms. See *Figure 12.4*:

Figure 12.4 – Change Layer mode to Color, and paint in the Skin Tone on the vector mask

Now that we've colorized the skin, let's move on to the next area.

Colorizing the eyes

Here are the steps for colorizing the eyes:

1. Next, go to **New Adjustment Layer| Color Fill** to create a new Layer. Rename it **Eye Color** and zoom in on the eyes.

2. Next, click on the white mask and press *Ctrl + L* to convert it to black.

3. Change layer mode to **Color**.

4. *Double-click* on the **Eye Color** Layer and adjust the color to dark brown for the eye color.

5. Begin painting the eyes; adjust the Layer **Opacity** accordingly. It may vary depending on the eye color and lighting situation. See *Figure 12.5*:

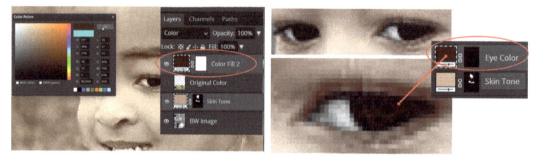

Figure 12.5 – Create a New Adjustment Layer, Color Fill, and paint the eyes

Now that we've colorized the eyes, let's move on to the next area.

Colorizing the lips

Here are the steps for colorizing the lips:

1. Go to **New Adjustment Layer | Color Fill** to create a new Layer. Rename it **Lip Color** and zoom in on the lips.

2. Next, click on the white mask and press *Ctrl + L* to convert it to black (to reveal the image).

3. Next, double-click the **Lip Color** layer to open the **Color Picker** window.

4. Reveal the original color photo reference, select the **Eyedropper** Tool, and sample the lip color.

5. For the lips, you may need to experiment with the layer modes to get natural-looking lips.

6. Set the Layer mode to **Overlay** and click on the black Layer mask. Make sure the brush foreground color is white.

7. Next, lower the **Opacity** of the mask Layer to around **60%**, or adjust accordingly.

8. Zoom in and start painting the lips. After you paint the lips, you can experiment with the **Soft Light** Layer mode and other modes that look natural (see *Figure 12.6*):

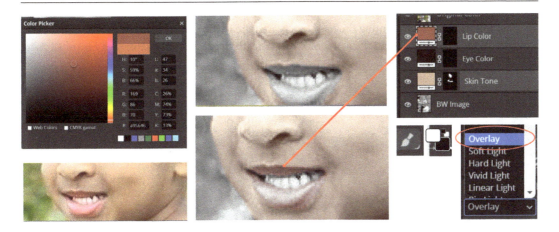

Figure 12.6 – Create a new adjustment layer, sample the lip color, and paint the lips

> **Important note**
>
> If the lips look too unnatural on the face, you may need to soften the edges of the lips to blend the color into the face using the **Feather** mode (*double-click* on the mask to activate).

Colorize the hair

Since the boy in the **BW Image** layer has a low haircut, some of the flesh tone (skin color) will show through the thin areas and gaps in the hair and around the edges and sides. We can go back and paint more skin color on the **Skin Tone** layer:

1. First, let's make a color fill adjustment layer, and rename the Layer **Hair Color**.

2. Next, press *Ctrl + L* to invert the mask to reveal the image.

3. Select the **Brush** tool, change the foreground color to white, switch the layer mode to **Color**, and set **Opacity** to around **60%**.

4.   Paint in the hair; after painting the hair, I increased the Layer **Opacity** to **85%** and revisited the **Skin Tone** mask to paint in the skin color underneath the **Hair Color** layer. See *Figure 12.7*:

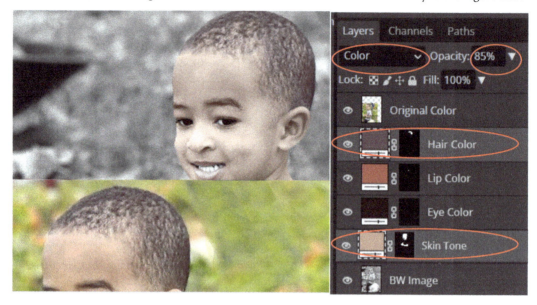

Figure 12.7 – Creating a new color fill adjustment layer and painting the hair

5.   Next, I switch the foreground color to black, select the **Hair Color** Layer, and lower the Brush tool's **Opacity** to about **25%** to remove and reduce some of the brown hair color and saturation simultaneously.

6.   I also did this for the **Skin Tone** layer. I reduced the Brush tool's **Opacity** to around **25%**, switched the foreground layer to white, and painted in the skin tone underneath the hair to give it a more natural transition between the hair and skin tone. See *Figure 12.8*:

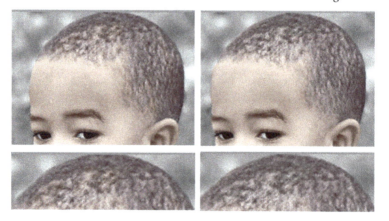

Figure 12.8 – Reducing brush Opacity and painting skin underneath the hair

We'll stop here with the hair color and skin tone for now. Let's next colorize the clothes and other areas of the photo before we invest too much time in the facial area.

> **Important note**
>
> Some things I have learned about painting portraits and illustrations are that it's best not to fully render one area of the painting too quickly. Instead, get the colors established in different parts of the painting, add the details, and work on the final revisions last. You can balance out what works and what's needed to achieve a great image with a built-up, layered approach.

Colorizing the clothes

Before we get started, note that I reduced the size of the original color photo with the **Move** tool and went to **Edit | Transform** to scale it down. This makes it easier to compare the new image while we paint. Now, let's colorize the clothes. See *Figure 12.9*:

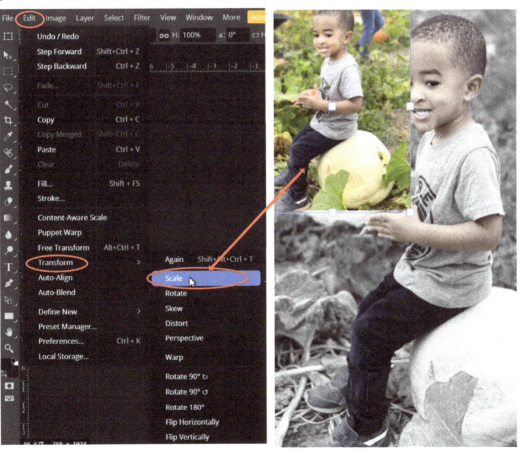

Figure 12.9 – Scale down the color photo for easy viewing while colorizing the BW Image layer

Since the shirt is already grey, we can begin the steps for colorizing the pants:

1.   Select the **Eyedropper** Tool to sample the pants in the original color photo. See *Figure 12.10*:

Figure 12.10 – Sample the pants' color from the original photo

2.   Create a **New Adjustment Layer** named **Color Fill 1**.

3.   Next, reduce the layer **Opacity** and rename the **Color Fill 1** layer to **Pants Color**.

4.   Click on the white mask and press *Ctrl + L* to invert it to black to reveal the image.

5.   Next, double-click the **Pants Color** Layer to open the **Color Picker** window.

6.   Begin painting on the **Pants Color** Mask. See *Figure 12.11*:

Figure 12.11 – Paint in the blue pants color

Now that we've colorized the pants, let's move on to the next area. Next, we will colorize the shoes:

1. Create a **New Adjustment Layer Color Fill** Layer and sample the navy color on the shoe with the **Eyedropper** Tool.

2. Next, reduce the Layer **Opacity** and rename the **Color Fill 1** Layer to **Shoe Color**.

3. Click on the white mask and press *Ctrl + L* to invert it to black to reveal the image.

4. Next, double-click the **Shoe Color** Layer to open the **Color Picker** window.

5. Reveal the original color photo, select the **Color Picker** tool, sample the Shoe Color, and begin painting on the Shoe Color mask. See *Figure 12.12*:

Figure 12.12 – Paint in the Blue Shoe Color

6. Create a separate **New Adjustment Layer** for the tan color accent of the shoes. Rename the Layer Shoe Accent Color. Repeat the process (*steps 1-6*) we did for the pants and shoe color. See *Figure 12.13*:

Figure 12.13 – Paint in the tan accent shoe color

Now we are ready to colorize the pumpkin that the boy is sitting on.

Colorizing the pumpkin

Let's use the following steps to colorize the pumpkin:

1. Create a **New Adjustment Layer** named **Color Fill 1**. Rename it **Pumpkin Color** and reduce layer **Opacity** to **60-70%**.

2. Next, click on the white mask and press *Ctrl + L* to convert it to black to reveal the image.

3. Next, double-click the **Pumpkin Color** Layer to open the **Color Picker** window.

4. Reveal the original color photo, select the **Eyedropper** Tool, and sample the pumpkin color.

5. Change the Layer mode to **Color**, select the **Brush** tool, make sure the foreground color is white, and paint the pumpkin. See *Figure 12.14*:

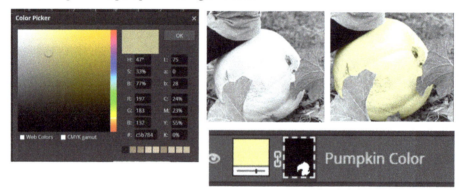

Figure 12.14 – Paint in the pumpkin

Now that we've painted the pumpkin, let's move on to the next area of the photo.

Colorizing the background

Now we are ready to paint the background:

1. Create a **New Adjustment Layer** named **Color Fill 1**. Rename it **Background Color**.

2. Next, click on the white mask and press *Ctrl + L* to convert it to black to reveal the image.

3. Next, double-click the **Background Color** Layer to open the **Color Picker** window.

4. Reveal the original color photo, select the **Eyedropper** Tool, and sample the green background color.

5. Switch the layer mode to **Overlay**, and reduce layer **Opacity** to **55%**.

6. Paint in the background color. See *Figure 12.15*:

Figure 12.15 – Colorize the background

Next, let's see how well we've painted in all of the masked layers thus far, by combining them all into one new mask.

Let's check how well we've colorized each section of the black and white image:

1. First, right-click the **Background Color** Layer and select **Delete Raster Mask**. See *Figure 12.16*:

Figure 12.16 – Delete Raster Mask

2. Next, we will make selections around all the mask layers to create one new mask for the next steps.

3. To begin, press the *Ctrl* key and click on the **Pumpkin Color** layer mask. This will make an active selection.

4. Next, press *Ctrl* and click each of the Mask Layers: **Shoe Accent Color, Shoe Color, Pants Color, Hair Color, Skin Color, Lips Color, Eyes Color** to activate them all as selections.

5. Now click on the **Background Color** layer. See *Figure 12.17*:

Figure 12.17 – Delete Raster Mask

6. Go to the **Layer** menu and select **Raster Mask** and then **Add (Hide All)**. See *Figure 12.18*:

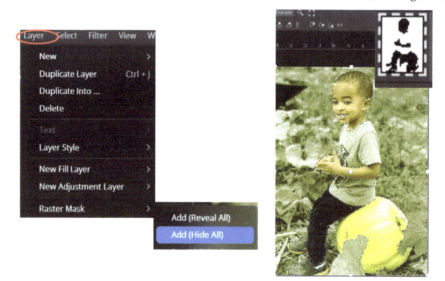

Figure 12.18 – Raster Mask (Hide All)

When we combine the mask layer selections, we will only have the **Background Color** Layer revealed, so that we can view our work, make edits, and clean up any spots that were missed in a black and white format.

7. Next, select the **Brush** tool and change the foreground color to black. Now, paint in black over the missed areas revealed in White. See *Figure 12.19*:

Figure 12.19 – Black and white mode to clean up missed spots in the image

Now that we've cleaned up the layers, let's move on to the next area.

8. Next, I decided to make a **New Adjustment Layer Color Fill** mask for the -Shirt Color because I didn't want the green background color bleeding through the grey shirt; and also, so that I could make the shirt a different color besides grey. This will make the colorized photo pop more. See *Figure 12.20*:

Figure 12.20 – Add a new adjustment layer color fill mask to the grey shirt

Now that we've colorized the shirt, let's move on to the next area.

9. After making the new adjustment color fill for the shirt, I revisited the hair color and skin tone saturation. I felt it needed more adjustments and refinements on the skin tone, eyes, and lips.

   The skin tone and lips needed warmer colors, such as a hint of red and orange to make the boy look more life-like. The warmth also provides a contrast against the cool colors of the blue t-shirt and green grass.

10. I decided to duplicate the **Skin Tone** and **Lip Color** Layers with the boy on the right.

11. *Right-click* the mouse over the **Skin Tone** Layer to *duplicate* the Layer, and repeat the process for the **Lip Color** Layer. The results turned out exceptionally well for me. You can see how the skin tone and lips of the boy are a little warmer and more intense than in the image on the left. See *Figure 12.21*:

Figure 12.21 – Revisit the lip color and skin tone and make additional adjustments

12. Next, I created a **Curve Adjustment** layer, **Curves 1**, to adjust the image contrast. See *Figure 12.22*:

Figure 12.22 – Make curve adjustments

13. I needed to add some color to the corner of the eye. I sampled the lip color with the *Eyedropper* tool and colorized the corner (tear duct) area of the eye on the **Lip Color** layer mask. See *Figure 12.23*:

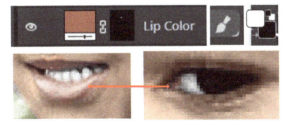

Figure 12.23 – Sample the lip color to paint the tear duct in the eye

It's time to colorize the pumpkin in the basket to the left of the boy's face.

14. I created a **New Layer Adjustment Color Fill** layer titled **Small Pumpkins** to paint the pumpkin in the basket with a mask layer.

> **Important note**
>
> The pumpkin could also be painted orange with the Soft Round Brush tool, as we will do in *step 15*.

15. I revisited the background to add some hints of orange color to portray pumpkins on the pumpkin patch.

     This was done with the **Round Soft Brush** tool. Set the Brush tool's **Opacity** to **21%**, and Brush size to **45%**.

16. Next, I created a default **New Layer** titled **Small Pumpkins 2** and painted the orange color on the canvas. You can see the brush marks made to paint them on a *transparent* layer. See *Figure 12.24*:

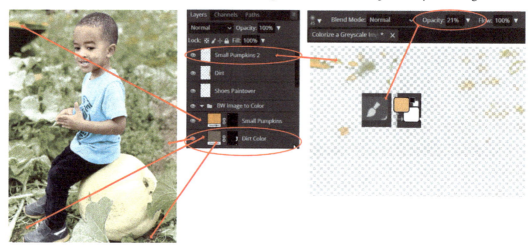

Figure 12.24 – Use the Soft Round Brush to paint orange pumpkin shapes

Now we have finished our new colorized image. You don't want to overwork the image. Remember, the goal is not to make an exact copy of the photograph. That would be impossible. We just need to make a balanced image and have fun doing it. See the final image in *Figure 12.25*:

Figure 12.25 – Final colorized image

That sums up how to colorize a black and white image. Let's move on to the next section to expand our learning of more color techniques.

Replacing a particular color

This section will go over a simple but powerful feature that allows you to change a particular color in your image. One thing to be aware of before we start is that this is a permanent change. I recommend backing up a copy of your image file and also making duplicate layer copies as backups, as shown in *Figure 12.26*:

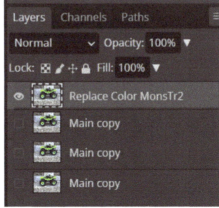

Figure 12.26 – Make duplicate layers of the image

Let's begin our work on replacing a particular color:

1. Locate the monster truck photo in the `Unlock Your Creativity Resources` folder from this chapter.

2. Go to the **Image** menu | **Adjustments** | **Replace Color**.

3. You will see a small, pink-colored button, just below the **Fuzziness** slider.

4. Double-click on the pink button to launch the **Color Picker** window.

5. Select the **Eyedropper** tool to sample the truck color and use the **color slider** to change the color. See *Figure 12.27*:

Figure 12.27 – Double-click the pink button, Eyedropper tool, and sample the truck color

6. Next, adjust the *sliders* for **Hue**, **Saturation**, **Lightness**, and **Fuzziness** to see the extreme, or subtle, color changes you can make. In the first example, I moved the **Fuzziness** and **Hue** sliders to the far right, which created a blue-purple color truck. See *Figure 12.28*:

Figure 12.28 – Adjust the Hue, Saturation, Lightness, and Fuzziness sliders

7. Next, here are two more examples of the same truck in different colors. The example on the right shows how reducing the **Fuzziness** can create a two-tone color blend seamlessly. See *Figure 12.29*:

Figure 12.29 – Increased Fuzziness (Left), Reduced Fuzziness (Right)

> **Important note**
>
> You can make duplicate layers of the original image to experiment and save different color variations.
>
> Making lots of duplicate copies of your original image layer before using the **Replace Color** feature will come in handy when you need to match or harmonize the colors of an object, or in this case, a toy truck for an advertisement design with a specific color palette; or you may need to present a number of color schemes to a client, and so on.

You can see more color variations of the truck with different settings on the **Fuzziness**, **Hue**, **Saturation**, and **Lightness** in the examples in *Figure 12.30*:

Figure 12.30 – More color variations

Now that we understand how to replace a particular color, let's move on to explore using the Color Replacement brush tool.

Using the color replacement brush tool

The **Color Replacement** brush tool is used for just that, replacing colors. It is a useful technique that you can do without having to create layer masks or complex selections. Yet, it's not at a silver-bullet solution for replacing colors since it's permanent and destructive. But it has its strengths, and can save a lot of time when used for certain projects.

Let's take a quick dive into using the **Color Replacement** brush tool to change the red color on the snowman's hat:

1. Locate the snowman photo in the Unlock Your Creativity Resources folder from this chapter.

2. To begin with, duplicate the **Snowman Original** layer.

3. Rename the duplicate layer to **Select + Change Color** for this example.

4. Next, select the **Color Replacement** brush tool. See *Figure 12.31*:

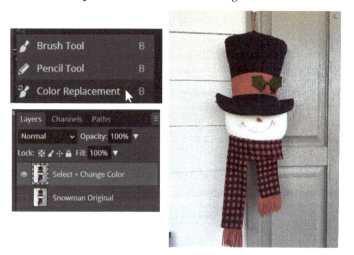

Figure 12.31 – Select the Color Replacement brush tool

5. Select the **Hard Round Brush** tool and set the brush size to **80px**.

6. Start painting the red band around the hat. Avoid the black area of the hat as best as possible. If you do paint on the black area, simply select the **Eraser** tool and erase any green color. See *Figure 12.32*:

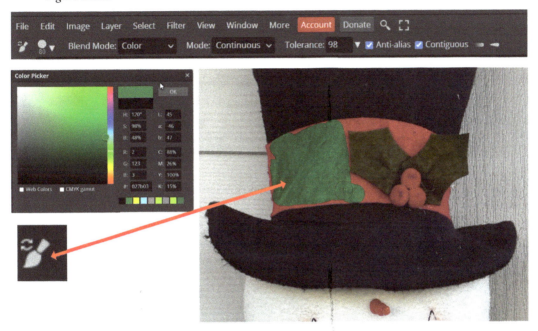

Figure 12.32 – Use the Hard Round Brush tool to paint in the green

Once the red band is completely recolored in green, make sure to check the edges along the hat for any missed bits of color or odd color changes.

> **Important note**
>
> Notice how the **Color Replacement** brush tool preserves the original texture of the hat's material, whereas the default Round Brush paints over the hat with the texture of the brush paint (see *Figure 12.33*).

Figure 12.33 – The Color Replacement brush preserves the original texture

7. Revisiting the hat, there were spots of red that I missed and an off-green color around the edges of the hat that stood out. To fix this, I reduced the size of the **Color Replacement** brush and painted over the parts of red that were missed.

8. To fix the off green color along the edges, I made a new layer, selected the **Soft Round Brush**, set the brush size to **15px** and **Opacity** to **20%**, and began painting over the green edges on the new layer. See *Figure 12.34*:

Figure 12.34 – Paint along the edges using the Soft Round Brush tool

9. The following figure shows a closeup of the snowman's red band replaced with green using the **Color Replacement** brush and some light touch-up work with the **Soft Round Brush** tool:

Figure 12.35 – The touch-up work finished

10. You can see the original version and the new version of the hat side by side in *Figure 12.36*:

Figure 12.36 – Original version versus the new version

That just about completes this section on using the **Color Replacement** brush tool to replace a color. Now that we understand how it is used in projects, we are ready to move on to the next section on modifying channels.

Modifying channels

In *Chapter 5*, we briefly covered working with channels in the context of making selections and masks with them. In this section, we will dive a little further into channels, exploring ways to modify them with the **Channel Mixer** and see how we can mix and create six specific colors using this approach.

This will also be a great exercise to expand your repertoire for enhancing images, design elements, and objects for your design work. The more you experiment with channels, the more confident you will be to try out and learn the technical side of channels for professional services and tasks such as color separation and 4-color processing (CMYK) for printing, decorated apparel, and for publications of books and magazines.

Let's begin our exploration of modifying channels:

1. To get started, create a new document of **1920 x 900 px**. Alternatively, you can go to **Templates | Screen**, select **Full HD 1920 x 1080 px**, then change it to **1920 x 900 px**, and change the **DPI** from **72 px** to **150 px** for a slightly higher resolution document.

   You will need three different images of the same figure or object. In this example, I'll be using a 3D portrait model of a woman in three different angles.

2. I was able to export the 3D model as a PNG file with a transparent background, so I didn't need to remove the model from the background using the selection tools in Photopea.

Important note

I used a 3D model included in the Character Creator 3 software. You have to purchase the software to use the model royalty free. You can purchase the software at (`https://www.reallusion.com`). I recommend using a similar free 3D program such as Daz Studio instead. You can access it at `https://www.daz3d.com/`.

---

3. Drag all three images onto the document and keep the **Background** Layer setting default **White**.

4. Next, rename your layers, starting from left to right, since we will be working in that order: **Fig 1 Left**, **Fig 2 Mid**, and **Fig 3 Right**. See *Figure 12.37*:

Figure 12.37 – Place the three separate images on the document and rename them

5. Next, make duplicate layers of each of the images and place them below/behind the original images.

6. Nudge or move each of the copies to the left or right, leaving a third of the copy image visible. See *Figure 12.38*:

Figure 12.38 – Move or nudge the duplicate layers to the left or right to make them visible

Now that we've set up the three 3D models and one copy of each model, let's move on to the next section and see how we can create three basic channels with the models.

Creating the three basic channels – Red, Green, and Blue (RGB)

As we know, red, green, and blue are the primary colors on the color wheel, and RGB is the default color mode for documents in Photopea (and all other image editing and graphics software that displays colors).

Let's create and simulate the primary colors on the 3D models beginning with the color red.

Creating the red channel

Let's begin creating the red channel:

1. Select the first image on the left, **Fig 1 Left**.

2. Go to **Image | Adjustments | Channel Mixer…**.

3. We will use the **Channel Mixer…** to modify each specific color.

4. Check which default channel color is active in the top left of the **Channel Mixer** window. Select the **Red** if it is not present.

5. Slide the **Red** slider over to the right, and slide the **Total** to the right, until you get a red color overlayed on **Fig 1 Left**. See *Figure 12.39*:

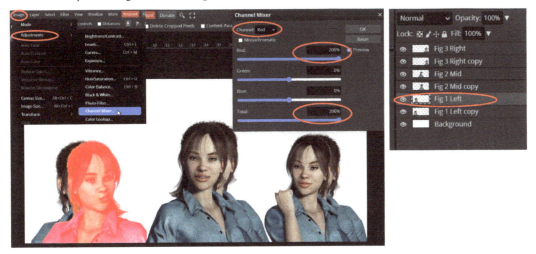

Figure 12.39 – Change Fig 1 Left to red with the Channel Mixer

Now that we've created the red channel, let's move on to the next one.

Creating the green channel

Let's work on the green channel:

1. Select the **Fig 2 Mid** model.

2. Go to **Image | Adjustments | Channel Mixer…**.

3. Switch the **Red** option to **Green** in the top left of the **Channel Mixer** window.

4. Next, slide the **Green** slider over to the right to **200**, and slide the **Total** slider to the right to **200**, until you get a green color overlayed on the **Fig 2 Mid** image. See *Figure 12.40*:

Figure 12.40 – Change Fig 2 Mid to green with the Channel Mixer

Now that we've created the green channel, let's move on to the next one.

Creating the blue channel

Use the blue channel as follows:

1. Select the **Fig 3 Right** model on the right.

2. Go to **Image | Adjustments | Channel Mixer…**.

3. Switch the color to **Blue** at the top of the **Channel Mixer** window.

4. Slide the **Blue** slider over to the right to **200**, and slide the **Total** slider to the right to **200**, until you get a blue color overlayed on the **Fig 3 Right** image.

5. Now, I recommend going to **Save as PSD** and saving a new copy of the file. This is just in case you lose any data, make a mistake, or experience file corruption (believe me, it happens). See *Figure 12.41*:

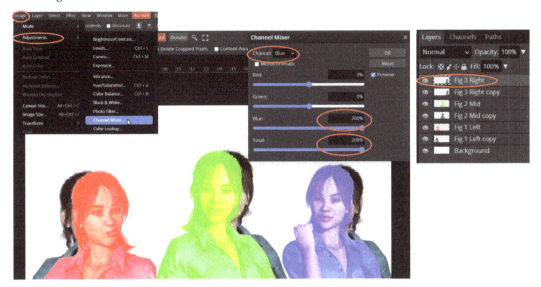

Figure 12.41 – Change Fig 3 Left to blue with the Channel Mixer

Now we can move on to the second part of this exercise.

Creating cyan, magenta, and yellow (secondary colors)

We will create the cyan, magenta, and yellow tones using the duplicated image layers.

Cyan channel

Let's begin with the cyan channel:

1. Select the **Fig 1 Left copy** image on the left.

2. Go to **Image | Adjustments | Channel Mixer….**

3. To create cyan, we will use a combination of mixing the main colors, green, and blue to mix the secondary colors.

4. Select **Green** in the top left of the **Channel Mixer** window, slide **Green** to **200%**, and the **Total** to **200%** (do not click **OK** yet).

5. Next, select **Blue**, in the top left of the **Channel Mixer**, and slide the **Blue** to **200%**, and the **Total** to 200%, and you will see a light bright blue color. Click **OK**. See *Figure 12.42*:

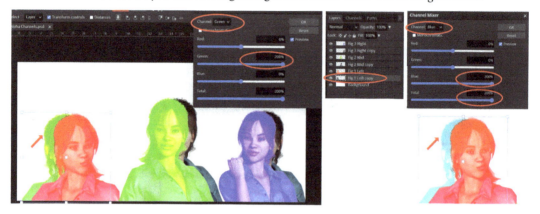

Figure 12.42 – Change Fig 1 Left copy to Cyan with the Channel Mixer

Now that we've created the cyan channel, let's move on to the next one.

Magenta channel

Let's begin work on creating the magenta channel. Magenta is a combination of red and blue:

1. Select the **Fig 2 Mid copy** image layer.

2. Go to **Image | Adjustments | Channel Mixer…**.

3. Select **Red** in the top left and slide **Red** to 200% and the **Total** slider to 200% (do not click **OK** yet).

4. Next, switch the **Red** channel option to **Blue**.

5. Slide the **Blue** slider to **200%**, and the **Total** slider to **200%**, and then click **OK** to create the magenta color. See *Figure 12.43*:

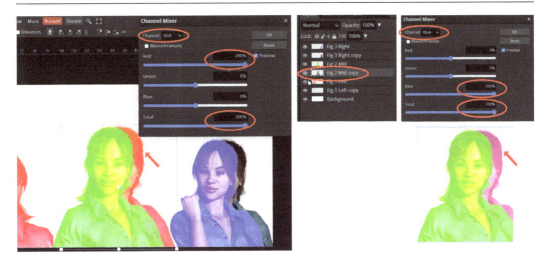

Figure 12.43 – Change Fig 2 Mid copy to magenta with the Channel Mixer

Now that we've created the magenta channel, let's move on to the next one.

Yellow channel

Let's set up the yellow channel. Yellow requires a mix of red and green:

1. Select the **Fig 3 Right copy** image on the right.
2. Go to **Image | Adjustments | Channel Mixer…**.
3. Select the **Red** channel option at the top of the **Channel Mixer** window.
4. Slide the **Red** slider to the maximum **200%**, and also the **Total** slider to **200%** (do not click **OK** yet).
5. Switch the **Red** option to **Green**.
6. Slide **Green** to **200%**, and the **Total** slider to **200%**.

7. Click **OK** to see the yellow overlay:

Figure 12.44 – Change Fig 3 Right copy to yellow with the Channel Mixer

It is harder to see the features and details with the yellow channel compared to the other channel layers, but they are there.

Yellow is the last color needed to create the six specific colors we wanted using channels.

Next, we will blend and modify some new colors from the six colors (the primary colors of red, green, and blue and the secondary colors of magenta, cyan, and yellow) we just created.

Creating a blend of colors

Let's create a blend of colors as follows:

1. Let's begin with the **Fig 1 Left** Layer, changing the layer mode on the primary color layers first.

2. Select the **Fig 1 Left** image on the right (the red image), and change the layer mode, located under the **Layers** tab, from **Normal** to **Multiply**. You can see how playing with the layer modes lets us blend the red and blue colors back to an almost natural skin tone where the two layers intersect. See *Figure 12.45*:

Figure 12.45 – Modify and blend the colors by changing the layer mode

3. Next, repeat the process by selecting **Fig 2 Mid** (the green image). Switch the **Normal** layer mode to **Multiply**. This will also create a new blend of colors that bring out the features of the face.

4. Now we are at the last blend. Select **Fig 3 Right** (the blue image). Switch the layer mode to **Multiply** to create the new color blend.

5. Notice how the yellow has blended with the blue and made the details in the face more visible. If you align the two layers, they will create a unique blend of colors. You've now seen how you can combine two color channels to create a unique-looking image:

Figure 12.46 – Modify and blend the colors by changing the layer mode to Multiply

6.   You can move the images closer to each other for clearer images of the subject. See *Figure 12.47*:

Figure 12.47 – Move the images closer together to see the facial features better

7.   You can go a step further and change each of the layer modes for the cyan, magenta, and yellow images to **Multiply** as well to see how the blends interact. If you have time, experiment with some of the other layer modes and see how the colors blend and interact. See *Figure 12.48*:

Figure 12.48 – Try some other layer modes for a unique look

That covers this section on how to modify channels. We learned another way to mix and assign specific colors using the **Channel Mixer**, and create color blends with three individual images for which we made duplicate layers, mixing primary and secondary colors.

Now let's summarize the chapter before we move on to *Chapter 13, Bonus: How to Draw and Paint a Figure and Character*.

Summary

This chapter was the second part touching on color. We learned some advanced color techniques to further our understanding not only of Photopea, but also of color theory more broadly that can be applied to just about any image editing program.

We learned how to colorize a black and white image using **Color Fill** layers, masks, layer blend modes, and other non-destructive adjustment layers on our images. In the process, we also learned how to convert a color image to black and white using the **Image Adjustments** feature.

Next, we learned how to replace a particular color using the **Eyedropper** tool in conjunction with the **Color Picker** and **Image Adjustments** windows to replace colors very quickly.

In addition to that, we learned a different approach to replacing color using the **Color Replacement** brush. We discovered how to replace a color by adding brush strokes over an object or a section of an image, similar to using a standard Round or Soft Brush tool, without the need for layer masks or making complex selections with the various selection tools, and better yet, keeping the initial textures and details intact.

Via the last section on how to modify channels, we gained a better visual understanding of the relationships between the primary colors (red, blue, and green) and the secondary colors (cyan, magenta, and yellow). We learned how to assign a specific color to individual images in a single document using the **Channel Mixer** window to mix primary and secondary colors with various layer blend modes.

That concludes *Chapter 12, Advanced Color Techniques*. It's time to move on to *Chapter 13, Bonus: How to Draw and Paint a Figure and Character*.

Bonus: How to Draw and Paint a Figure and Character

Gaining a solid foundation in drawing and painting can take your image editing skills to new levels. It can open a wider range of creative possibilities and potential job opportunities. On a personal level, drawing and painting teach you how to analyze, observe, and recognize what you see in front of you, rather than just making assumptions about what you see in front of you when you are drawing and studying from live subjects.

In this chapter, we will draw a figure and a character from start to finish. The character I created is named **Mother Earth**. She is based on themes relating to conservation, the impact of global warming on the environment, and so on.

Using a character like Mother Earth will help us learn how to come up with ideas for a character, organize our ideas and drawings, and learn the basics of mixing and rendering details with paint.

We will begin the project with rough sketches, followed by selecting the best composition to polish up the final drawing for a grayscale value study, before finishing up in full color.

For this chapter, we will cover the following topics:

- Getting started with drawing and painting
- Sketching and developing a character design
- Grayscale study and adding color to your character

Getting started with drawing and painting

People often get intimidated when it comes to drawing, often because they are looking at the fine details of a finished drawing or painting. Essentially, drawing consists of simple lines, curves, and shapes that combine to make life forms (animals and human figures) and man-made objects.

Technology today makes it easier to dive into drawing and painting. Image editing and drawing software such as Photopea, Photoshop, and Painter can be used with drawing tablets. These tools allow you to draw using layers, edit content, undo actions, and so on. There are no worries about wasting paper, paint, and canvases.

In this section, we will look into a few different ways to begin a drawing, as well as tips for beginners on ways to practice using a drawing tablet.

Getting comfortable with drawing and painting

There are a number of ways to get comfortable with drawing. Firstly, find images and subject matter that interest you; it will help you to stay consistent and willing to practice regularly.

Start out with simple subjects before you try drawing complex forms such as the human figure. Some good examples of simple subjects are rocks, insects, birds, and fish.

A drawing tablet would be extremely beneficial in getting you comfortable with drawing complex subjects. For example, you can bring images into Photopea, lower the layer opacity, create a new layer, and try tracing over the images with a brush tool. Don't rush to trace the images: take your time to gain control and coordination.

Sketching an eagle

Let's take a look at an example of sketching an eagle. This is a great way to find out which brushes work for you:

1. Select **Hard Mechanical Default Brush** at **Brush Size 3px**, and set **Opacity** to **20%**; this will give you a sturdy, loose feel when sketching.

2. Take a moment to observe the eagle. Look for the simple shapes that make up the eagle's head, beak, wings, and body.

3. You can drag the ruler guides to give you an idea of the subject's landmarks and proportions. You can draw lines and sketch marks to act as guides for measuring the wings, body, and head. You can also hold your tablet pen to the monitor and measure the eagle as you would use a tape measurer. See *Figure 13.1*:

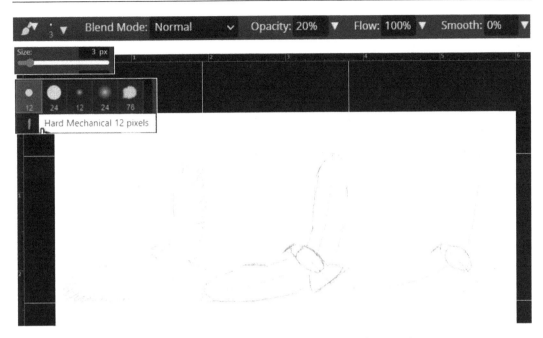

Figure 13.1 – Sketching an eagle using simple shapes, lines, and curves

For other examples of simple subjects to help you get comfortable with drawing, both frogs and turtles are made up of ovals and simple curves. I always start out with simple shapes and add the details last.

The following portrait of a man is an example of tracing over a photo to help aid in hand-eye coordination and getting comfortable with drawing using a tablet:

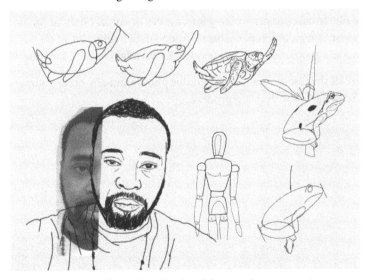

Figure 13.2 – Sketching simple subjects and tracing a photo

Now that we've explored how to get comfortable with drawing by starting with simple shapes and lines, let's dive into the next section.

Sketching and developing a character design

Now that we're aware of simple methods to approach sketching, let's look at ways to develop a figure drawing into a character design. This should include brainstorming, and writing ideas for your character's personality, physical traits, skills, interests, gifts, special powers, costume, and so on.

> **Important note**
>
> Before we begin this section, note that I was halfway through drawing the character design when I noticed a lot of lag and unresponsiveness while painting in small areas of the face. The paint was either not flowing from the brush or skipping spots before it laid down paint.
>
> I tried different brushes; I thought it was my pen tablet malfunctioning; so I changed nibs. I had no luck until I opened a different image editing program and saw that it was working just fine.
>
> So, I went to the **Image** menu and reduced the inches of the file from **8.5" x 11"** to **6" x 7.76"**, and the paint started to flow normally. Be mindful of file size, especially when you're using a lot of layers, as it will cause Photopea to drag.

Now that we understand some of the technical issues that can occur with large file sizes and layers, let's move on to the first step.

Finding photo references

The first step is finding photo references. If you can't come up with a character of your own, recreate a character that you like from a popular video game, movie, or comic book, or follow along with my character; give it a twist using a different costume, colors, nationality, and so on. Just be sure to find photos of human models, props, fashion, and textures to aid as a reference for your work.

Now that we've done all the research and gathered our photos and any other sources of inspiration, we can begin the next stage of the drawing process.

Drawing rough sketches and poses

Now we are ready to begin the sketching process, starting with finding the right pose:

1. In this example, I will sketch out three of my favorite poses out of all the photo references I found.

2. Select the default soft round brush to begin drawing loose mannequin sketches of the character in poses. See *Figure 13.3*:

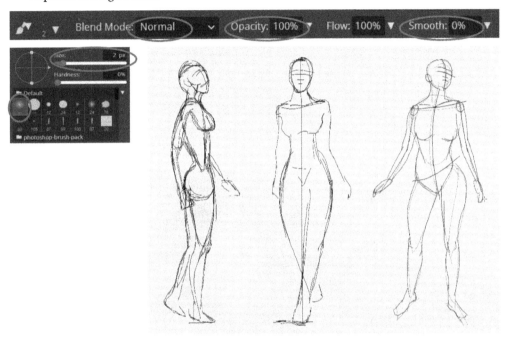

Figure 13.3 – Rough mannequin sketch poses

3. Create a new layer and draw a more polished line drawing on top of our favorite loose mannequin pose.

4. I used the default round hard brush at **100%** opacity. We can use the polished **Rough Sketch 2** for the next stage: drawing and developing the costume. See *Figure 13.4*:

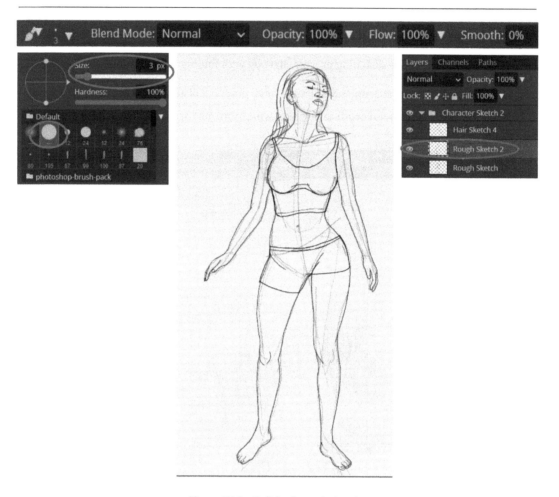

Figure 13.4 – Polished rough sketch

5. Go to **File | Save as PSD** and rename the document **Character Design Costumes**.

6. Now that the figure is drawn out, you can rename it **Final Fig Sketch 1**.

7. Next, duplicate the **Final Fig Sketch 1** layer a couple of times.

8. Rename the layers **Final Fig Sketch 2** and **Final Fig Sketch 3**, so that we can explore a few different costume designs for the character. See *Figure 13.5*:

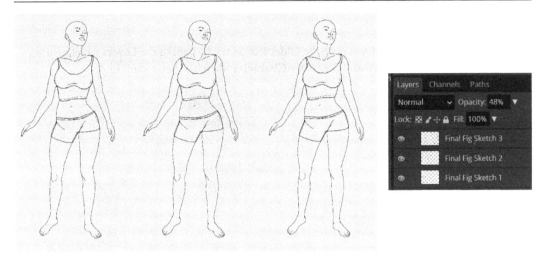

Figure 13.5 – Make duplicate layers of the final drawing

> **Important note**
>
> It's important to keep your layers organized, as it can get very messy and hard to manage if you don't name and arrange them in an easy-to-recognize order.

9. Next, create a new layer to sketch out different costume ideas on top of the first figure, **Final Fig Sketch 1**.

10. Create a separate layer, rename it **Dress Sketch 1**, and start sketching out the dress.

> **Important note**
>
> You may need to create two more separate layers to draw other versions of the dress to find the best style.

11. Next, create a separate layer, rename it **Hair Sketch 1**, and sketch out a hairstyle.

12. Next, create a separate layer, rename it **Accessories Sketch 1**, and sketch out the accessories.

13. I created a separate layer for the earrings in case I have to adjust the hair in front of the accessories as I progress in the painting.

14. Next, select all of the layers and click the folder icon at the bottom of the **Layers** panel to group them under one folder. Rename the folder **Character Sketch 1**. See *Figure 13.6*:

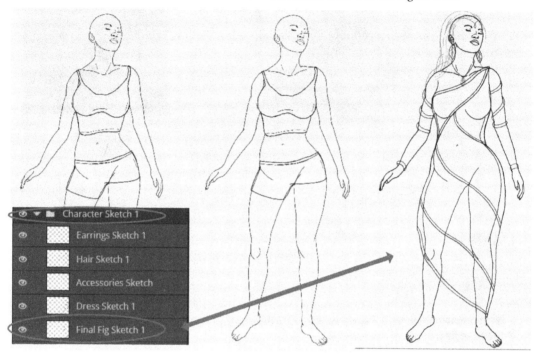

Figure 13.6 – Create separate layers for the dress, accessories, hair, and earrings

15. Duplicate the **Character Sketch 1** folder two more times.

16. Rename the second folder **Character Sketch 2** and the third folder **Character Sketch 3**.

17. Since the **Character Sketch 2** and **Character Sketch 3** folders have the same drawings as **Character Sketch 1**, I used parts of the drawings to continue to build new styles and erased parts of the sketches of the initial dress, accessories, and hair until I had created some different looks for each of them. See *Figure 13.7*:

Figure 13.7 – Create different costume styles for each figure

18. Next, select which character design concept you would like to paint. I decided to work with **Character Sketch 3** on the right.

19. Click **Save as PSD File**, rename the file **Grayscale Study**, and delete the other two **Character Sketch** folders in the **Layers** panel. (I have a backup of the characters from when I clicked **Save as PSD File**.)

20. Next, use **Save as PSD File** again and rename the file **Character Design Color**. That way, we have both files set up to do a grayscale study and get a rough idea of establishing contrast, shadows, and light sources with just three or four values.

Now that we've chosen the best costume design out of the three styles, let's move on to the next stage.

Grayscale study and adding color to your character

Before diving into adding color to our character design, it can be very beneficial to do a grayscale study of our character. We will only need three shades of gray to quickly capture the dark shadows, mid-tones, and brightest light or white colors and see whether our painting is balanced with good contrasting color values. Adding colors first may tempt you to get too detailed too quickly and dive into mixing, blending, and painting your character, only to find that the lighting and colors don't work.

Creating a grayscale study

Now we are ready to begin working on the grayscale study:

1. For this particular assignment, we will just use the default soft brush tool to paint a three-color image using **Black**, **Neutral Grey** (for mid-tones), and **White** (for lighter values). If the character design displays well with good contrast and lighting in black and white, it will make it easier to paint the character in color.

2. Next, set **Opacity** to around **25%** and **Smoothness** to **0%**.

3. I set the lighting to come from the upper right and the shadows to flow to the left side of her. This lighting is purely from my imagination so it may not be perfect.

4. You can set up a light source with some lamps or photography lighting in your room to figure out how you want to have your lighting, or set up a 3D model in a 3D program if you're comfortable using software for it.

> **Important note**
>
> Another approach you could take to adding color is to fully paint the character in grayscale (black and white), along with all the fine details, and then colorize the character similar to the way we added color to a black-and-white image in *Chapter 12, Advanced Color Techniques*.

I will just show the rough black-and-white character painted so that we can make room for the color process in the next section. See *Figure 13.8*:

Figure 13.8 – Grayscale study of the character as a guide to rendering in color

Now that we understand how beneficial it is to paint grayscale studies before taking on color, let's move on to the next stage of character design development.

Adding color to your character

Now it's time to open the **Character Design Color** file we saved in the previous section:

1. We will begin by adding flat colors to the figure.

2. Create a new layer underneath the **Final Fig Sketch 1** layer and rename it `Final Fig Color` **Flat**.

3. I sampled a woman's skin tone from some photos I found, to give me a quick start. (I cannot show the photos due to copyrights).

4. Select the lasso tool and make a rough selection around the original **Final Fig Sketch 1**.

5. Set **Edit Fill Color** for the skin tone.

6. Select the eraser tool and erase the extra paint around the figure.

7. If you accidentally erase part of the figure, select the hard round brush at **100%** opacity and paint the skin tone back in.

8. Use the eyedropper tool if you lose the colors in the color palette; see *Figure 13.9*:

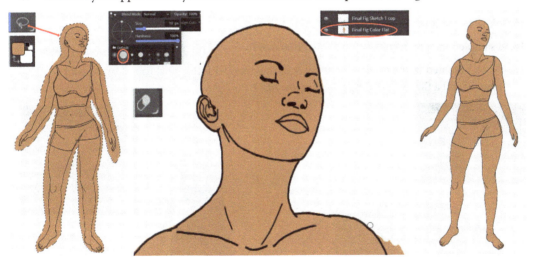

Figure 13.9 – Create a new layer and add a flat color to the figure

9. Next, create a new layer underneath the **Dress Sketch 1** layer and rename it **Dress Color Flat**.

10. Now, create a new layer above the **Dress Color Flat** layer and rename it **Dress Accent Color Fl**.

11. Next, I used the eyedropper tool to sample a blue-colored dress from some photos I found to give me a quick start.

12. I selected the hard round brush at **100%** opacity and **Size** at **10px**.

13. I started painting along the edges of the dress first, then I increased the **Size** of the brush to about **45px** and painted in larger areas of the dress.

14. I repeated this process until the dress was filled completely. See *Figure 13.10*:

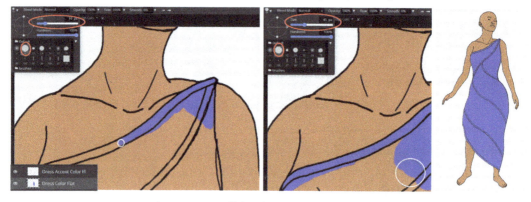

Figure 13.10 – Fill the dress with a blue flat color

15. Now we paint in the **Dress Accent Color Fl** Layer.

16. I sampled the green color from some photos of trees and plants with the eyedropper tool.

17. Next, I painted in the green accent color of the dress with the hard round brush tool at **100%** opacity and a **Size** value of **8px**. See *Figure 13.11*:

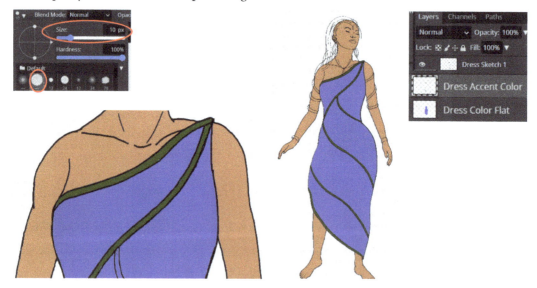

Figure 13.11 – Paint the Dress Accent Color with Flat layer green

18. Make a new layer below the **Accessories Sketch** Layer and rename it **Accessories Color Flat**.

19. Use the hard round brush tool at **100%** opacity and the **Size** value at **3px**, and paint in the same blue and green flat colors on the earrings, necklace, and arm bracelets.

20. Now we can make the **Accessories Sketch** and **Hair Sketch 1** Layers visible. She is coming along. See *Figure 13.12*:

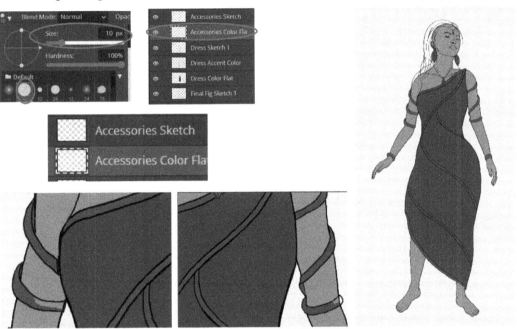

Figure 13.12 – Paint the Accessories Color Flat layer green

21. Next, create a new layer underneath the **Hair Sketch 1** layer and rename it **Hair Color**.

22. I sampled the hair color from hairstyle images of women online.

> **Important note**
>
> We need a soft brush since hair has a soft texture.
>
> If you painted the hair with a hard brush, it would look flat and heavy like a football helmet (as my life drawing mentor constantly emphasized).
>
> A hard brush would probably be fine for drawing flat, cartoon-type work.

Let's find a brush that will be good for painting the hair.

23. Go to **Window** | **Plugins** | **Brushes** and select **Art Brushes** by Dmitry Pavlov.

24. Click **Install**. Click on the brush tool and select **HGJart** from the **Photoshop Brush** pack. See *Figure 13.13*:

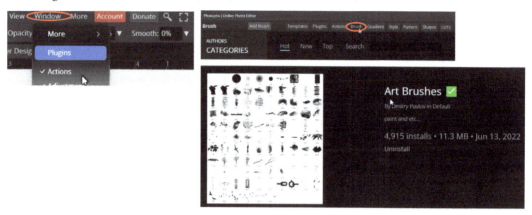

Figure 13.13 – Install the free Photoshop brushes by HGJart

25. Set the **Size** to **25px**, **Opacity** to **25%**, and **Smoothness** to **0%**.

   This is a *pressure-sensitive brush*, meaning the more pressure I apply to the pen tablet, the more paint will flow on the canvas. The less pressure I apply, the less paint will flow. It will have a softer look, which is what we need for the hair.

26. Begin painting in the direction of the hairstyle, leaving less paint near the edges of the hair. See *Figure 13.14*:

Figure 13.14 – Loosely paint in the hair color

> **Important note**
>
> We don't need to fully render the hair right now; we will continue working on the hair after we render the flat colors of the figure, dress, and accessories, and establish the light and shadows.

Painting in the color details

In this section, we will begin blending colors, and adding light and shadows over the flat colors, using clipping masks and the brush tools.

Before we get started, I decided to darken her skin tone:

1. Select the Move Tool, press and hold the *Ctrl* key, and click on the **Final Figure Color Flat** layer to make a selection.

2. Next, select the eyedropper tool to sample the skin tone.

3. Double-click on the foreground color to launch the **Color Picker** window.

4. Slide the **Color** to a darker brown color and press **OK**.

5. Go to **Edit | Fill** to update the skin. See *Figure 13.15*:

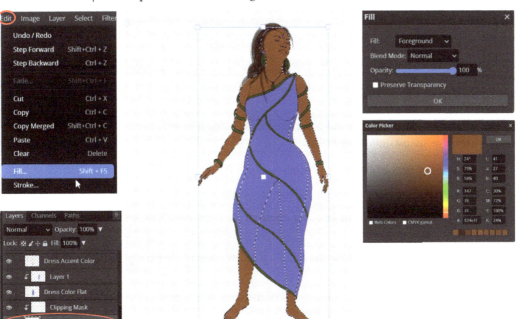

Figure 13.15 – Darken the skin tone

Now we can focus on painting in the details of the character.

6. I downloaded a free *Illustration Brush Set* by Matt Heath at Gumroad.com to paint the majority of the color. This was before I realized Photopea was lagging, so I tried some different brushes to see if that would solve the issue, which it didn't, but I gained new 20 brushes from this set.

    If you want to try out Matt Heath's Brushes, go to his site at the following link and add your email address to receive them: https://mattheath.gumroad.com/l/freepsdbrush?layout=profile.

7. Next, create a new layer above the **Final Fig Color Flat** layer, then right-click to make a clipping mask.

> **Important note**
>
> Creating a clipping mask for the Flat Color Layers will allow us to paint the skin tone without painting outside the lines or anywhere else on the figure.

8. Next, select the eyedropper tool to sample the **Final Fig Flat Color** skin tone.

9. Double-click on the foreground color to open **Color Picker**.

10. Adjust the slider to create a lighter brown color to paint on the clipping mask.

11. Next, go to the **Brushes** panel and select **MH 88 Brush 2 3**.

12. Keep the brush's **Opacity** value low, to gradually blend the colors together.

Mixing and blending colors

Let's look at an example of mixing and blending colors, using a brush to give you a better idea (you can follow along with just about any brush):

1. First, lay some flat paint strokes at **100%** opacity side by side.

2. Use the eyedropper tool to sample the light brown color.

3. Next, select the brush tool and make a single stroke over a small section of the darker brown. *Don't apply a lot of pressure.*

4. Now, select the eyedropper tool and sample the new color made from painting the light brown over the dark brown.

5. Paint a new brush stroke on the left side of the darker brown to see the subtle change as the colors blend into each other. See *Figure 13.16*:

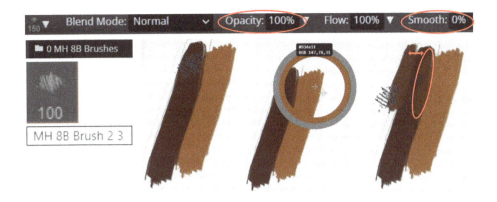

Figure 13.16 – Sample with the eyedropper and add a light paint stroke for softer blends of paint

6. Next, repeat the process to add and blend more colors together.

7. Use the eyedropper tool to sample colors.

8. This time, change **Opacity** to **20%**.

9. This will allow us to gradually add color on top of other colors to make smooth color blends. See *Figure 13.17*:

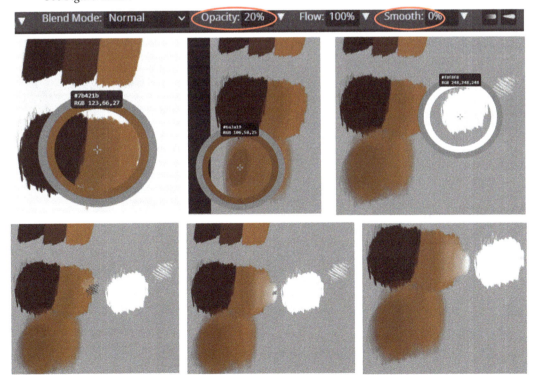

Figure 13.17 – Sample with the eyedropper and add light paint strokes for softer blends of paint

Now that we have a solid understanding of blending colors with the brush tool, let's move on to the next stage.

Blending colors throughout the character

In this stage of painting the character, you will see how blending in colors will add a sense of realism to the character. Let's begin the process:

> **Important note**
>
> If you haven't done so already, create a new layer above the **Final Fig Color Flat** layer. Right-click on the new layer and click **Add a Clipping Mask**.

1. Set the brush size to **20px**, **Opacity** to **20%**, and **Flow** and **Smoothness** to **0%**.

2. Use the eyedropper tool to sample the **Final Fig Color Flat** color.

3. Double-click on the foreground color to launch **Color Picker**.

4. Drag **Color Picker** down to a darker brown.

5. Begin painting the dark brown color on the **Clipping Mask** Layer, to give the face some form. See *Figure 13.18*:

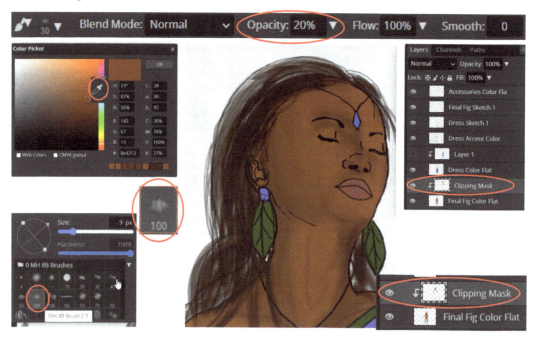

Figure 13.18 – Add a clipping mask to render the skin

6. Repeat the steps for painting the dark brown color on the arms and legs using the blending and mixing method in the previous section.

7. Next, use the eyedropper tool to sample a small area of the legs that has darker skin, and paint one or two strokes of that dark color over a mid-tone (or neutral skin area) to create a third color (a new in-between skin tone); it will be lighter than the darkest color but darker than the lightest skin color in that area. See *Figure 13.19*:

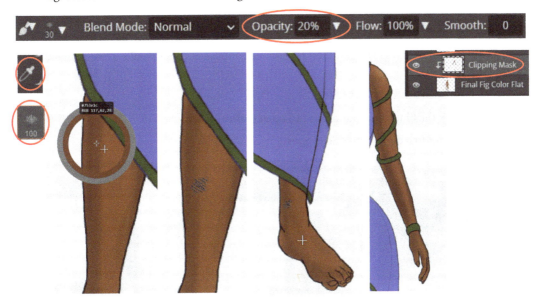

Figure 13.19 – Sample the skin tones to create a third color to blend

8. Continue adding the third color. Create a fourth color of lighter brown with **Color Picker**, and paint the lighter side where the light source is shining down on her from the upper right side (on the face and arms). See *Figure 13.20*:

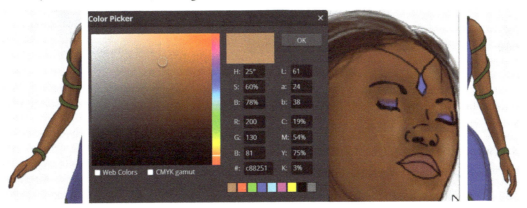

Figure 13.20 – Sample the skin tones to create a fourth color to blend

We can take a break from rendering the skin tone; it's time to paint the details for the dress and accessories. We'll have a better idea of how much detail to add to the skin tone and different areas of the character and costume after we get a good balance of color, contrast, brush variation, and values.

Painting the dress

I decided to make the flat color of the dress a darker blue:

1. Select the **Dress Color Flat** Layer so that the layer is active.

2. Now press and hold the *Ctrl* key, then click the left mouse button to make a selection around the dress.

3. Use **Color Picker** to sample the blue color of **Dress Color Flat**.

4. Double-click the blue foreground color to open **Color Picker**.

5. Drag the slider to a darker blue. See *Figure 13.21*:

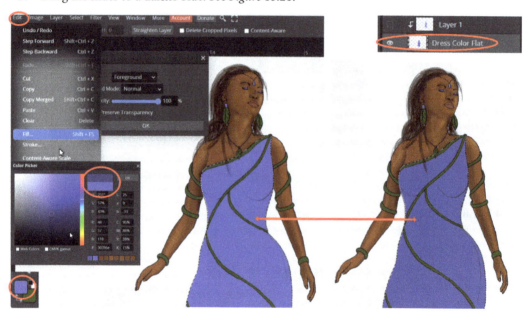

Figure 13.21 – Select the Dress Color Flat layer to darken the blue

6. Next, create a new Layer above the **Dress Color Flat** layer and right-click to make a clipping mask.

7. Change the brush to **MH Soft Round + noise**, set **Size** to **60px**, and set **Opacity** to **20%**.

8. Select the eyedropper tool to sample the blue dress.

9. Double-click on the blue foreground color to launch **Color Picker**.

10. Drag **Color Picker** to a slightly darker blue.

11. Press **OK**, and start painting some of the darker blue along the left side of her dress. See *Figure 13.22*:

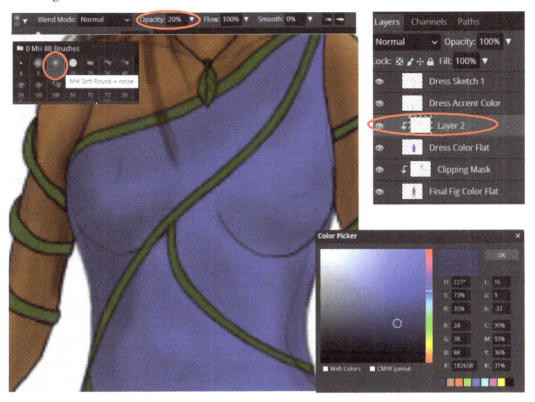

Figure 13.22 – Create a clipping mask and paint the dark blue shadows

12. Next, select the **Full Final Fig Sketch 1** Layer and move it above the **Dress Color Flat** Layer to give an idea of where to paint some of the imprints of the leg underneath the dress.

> **Important note**
>
> The **Full Final Fig Sketch 1** layer is a copy that has the entire outline of the lady's body without clothing. The **Final Fig Sketch 1** layer will eventually lose a lot of its outline; also erase some of the outline areas for the Paint Strokes to replace them.

13. Reduce **Opacity** to **20%** for **Full Final Fig Sketch 1** to act as a guide while we paint more of the dark blues and light blues on the dress around the anatomical form of the legs, chest, and hips underneath the dress. See *Figure 13.23*:

Figure 13.23 – Sample the blue to add darks and lights along the contours of the body

14. Next, create a new layer above the **Dress Accent Color** layer and make a clipping mask to paint the details over the green accent color.

Painting the accessories

Now it's time to paint in the details of the accessories:

1. Create a new layer above **Accessories Color Flat** and right-click to make a clipping mask.

2. Use the **MH Soft Round + noise** brush, set the **Size** to **60px**, and set **Opacity** at **20%**.

3. Paint in variations of dark green and light green. Repeat this process for the accessories. See *Figure 13.24*:

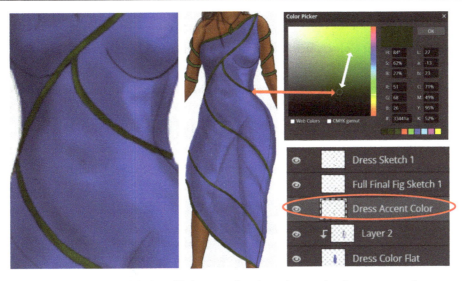

Figure 13.24 – Add dark and light green brush strokes on the dress accent color

Now that we've painted in the details for the accessories, let's observe the painting for a moment to see how the painting is coming along and see if there is anything we need to address.

Reduce some of the layers

After painting some of the details for the accessories, I feel we need to reduce some of the layers to make it easier to paint and reduce the file size:

1. Select the **Accessories Color Flat** Layer and the clipping mask.
2. Right-click, click on **Merge Layers** and rename the Layer **Accessories Color**. See *Figure 13.25*:

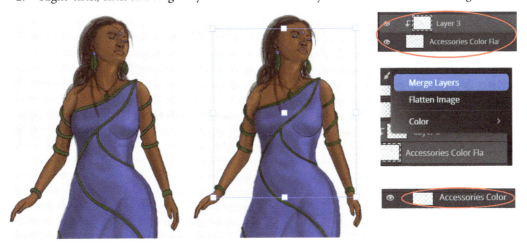

Figure 13.25 – Merge the Accessories Color Flat and Clipping Mask layers

3. Repeat this process for the other layers with clipping masks.

4. Select **Dress Accent Color**, place it above the **Clipping Mask** layer, right-click, and click on **Merge Layers**. Rename the Layer **Dress Accent Color**.

5. Select **Dress Color Flat**, place it above the **Clipping Mask** Layer, right-click, and click on **Merge Layers**. Rename the layer **Dress Color**.

6. Select **Final Fig Color Flat**, place it above the **Clipping Mask** layer, right-click, and click on **Merge Layers**. Rename the layer **Final Fig Color**.

7. Now each Layer that had a clipping mask was merged as a single Layer for each section, instead of two. See *Figure 13.26*:

Figure 13.26 – Each layer that had a clipping mask was merged as a single layer

Though we merged some of the layers, we can always create a new clipping mask when we're ready to paint in more details in each section of the character design.

Adding a background

Next, let's add an abstract-shaped background so our character doesn't look like she's floating on an empty screen. The background color will also help us decide on how to finish coloring and adding details to the hair:

1. Go to **Tool Box** and select the rectangle shape tool.

2. Click and hold the right mouse button and drag the mouse to the left or right to enlarge it as a backdrop.

3. Next, double-click **Fill** in the top menu and change the background color to a green earth tone. See *Figure 13.27*:

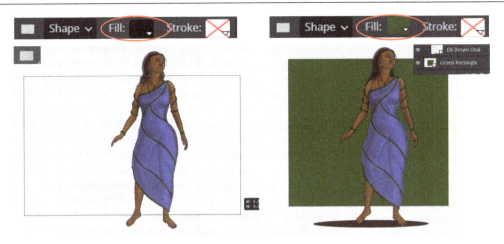

Figure 13.27 – Add a background with the Shape tool

4. Select the dark green color, reduce the **Opacity** for the layer to **60%**, and right-click over the layer to rasterize it (optional).

5. Select the ellipse shape tool, draw an ellipse underneath her feet, and change the color to dark brown.

    This reduces the intensity and saturation of the green in the background and allows the character to pop out of the canvas on the foreground. See *Figure 13.28*:

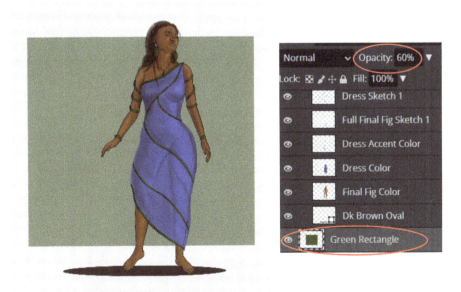

Figure 13.28 – Set the Lower Green Rectangle layer's opacity to 60%

Now that we've added a background, let's work on adding more details.

Touching up the figure

Now let's work on adding more details and style to the hair:

1. Select the eyedropper tool to sample the green background color and add it to the edges and inner areas of the hair with the **HGJart** brush.

2. Set the brush's **Opacity** value to **20%** and paint the green into the hair.

   This helps unify the painting with the character, hair, and background and softens the hair. See *Figure 13.29*:

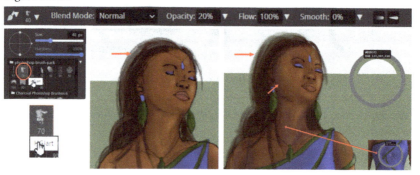

Figure 13.29 – Sample the green background color and add it to the hair

Let's go back and fine-tune the character design by painting in more details of color, shadows, highlights, and any other necessary adjustments beginning with the face and skin tone.

3. Select the eraser tool to erase a little bit of the edges by the right side of the eye, cheekbone, and chin to the figure on the left. Since she's slender, you will see more of the curving bone structure underneath the skin.

4. Next, paint the color back into the side of the face and add brighter skin tones and highlights. This will add more emphasis on her face and upper body. See *Figure 13.30*:

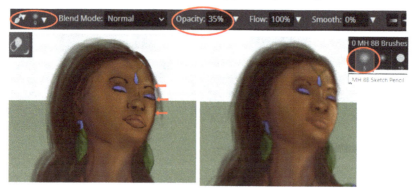

Figure 13.30 – Trim off some of the paint on the right side of the face and add highlights

Now that we've touched up the hair and added more details to the face, let's take a moment to observe the painting and see if any other improvements can be made.

Revisiting the dress and painting in more details

After observing the painting, I felt I should make the dress fit closer to the contours of her hips and legs. Let's make the corrections in a few steps:

1. Trim off some of the dress with the eraser tool. Erase some of her **Blue Dress Color**, **Dress Accent Color**, and **Dress Sketch 1** layers, so that the dress fits along the contours of her legs and hips. See *Figure 13.31*:

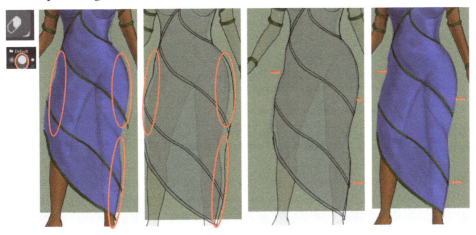

Figure 13.31 – Trim off some of the dress to fit the contours of the body

2. Next, I used the eyedropper tool to sample the blue color from the dress to paint her fingernails and toenails blue. See *Figure 13.32*:

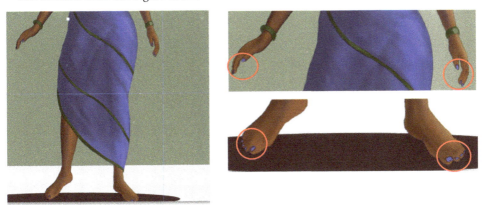

Figure 13.32 – Paint the fingernails and toenails blue

3. Next, I decided to duplicate the green background to try a different rectangular shape.

4. Afterward, I selected the default soft round brush, changed **Opacity** to **20%**, and added some soft white paint near her face and along the right side of her body. This gives a subtle glow effect and brings attention to her face. See *Figure 13.33*:

Figure 13.33 – Sample and add

5. I cropped the green square and added part of a plant I sketched as a painting in the background. I also softened the edges of the dress, arms, and legs, and added more highlights to her face, to strengthen the contrast between her and the background. I will consider this one finished; although I could keep on editing it, I don't want to overdo the demo. See *Figure 13.34*:

Figure 13.34 – Finished character design

Here is a close-up of the face. It creates a new composition that will work nicely as a portrait. See *Figure 13.35*:

Figure 13.35 – Finished character design cropped as a portrait composition

That concludes this section on touching up the hair, face, and painting in additional details of the character, adding a background, revisiting the dress, and painting in more details. We will summarize the chapter in the next section.

Summary

Chapter 13, Bonus: How to Draw and Paint a Figure and Character, is a bonus chapter that expands on other ways of using photo editing tools for drawing and painting.

We briefly demonstrated how those who may have only done photo editing can learn how to draw and paint if they are open-minded and learn how to see rather than assuming what they see.

From that point, we learned that there are a number of ways to get comfortable with drawing, such as starting out drawing very simple objects and subject matter and gradually working your way up to more complex human figures and objects.

We also touched on how technology and the drawing tablet make it easier to get into drawing, and how learning to trace can help with hand-eye coordination.

For the bulk of the chapter, we learned how to take rough sketches and ideas, and we developed them into a finished line drawing that will be the foundation for us to develop a character, by starting with the costume design and creating a rough grayscale rendering to use as a base to paint and add color, lighting, and composition details.

That concludes this chapter; in the next chapter, we will learn how to create a logo.

Bonus: How to Create a Logo

In this chapter, we will learn how to create a logo. We'll start with understanding what a logo is and how it is used, then progress to brainstorming, sketching out rough ideas, creating a logo in black and white, and then try adding color for a finished logo design.

By the end of this chapter, you'll have a better understanding of what a logo is and how to apply what you've learned in your designs. You'll also know how to build a design from conceptual ideas and select colors related to the mood and emotion of your logo design. Overall, you'll be able to create a well-designed logo for your brand or client.

In this chapter, we will cover the following topics:

- What is a logo?
- Sketching out rough ideas and brainstorming
- Creating a black and white logo
- Exploring colors and fonts

It's important to note, before we begin this chapter, that image editing programs such as Photopea and Photoshop aren't usually the first choice for designing logos, but can still be used to accomplish the final goal of a well-designed logo. Programs such as Adobe Illustrator, Inkscape, and CorelDraw are the top choices for creating and designing logos as they are vector-based programs. Vectors can be infinitely scaled, allowing the logo to be used in any number of size formats. Sizes can range from small stickers to billboards, and beyond. Photopea and Photoshop are, however, great for sketching out rough drawings and ideas for potential logo concepts.

Let's begin with the foundations of logos.

What is a logo?

Essentially, a **logo** is a symbol consisting of text and/or images, used to identify a brand. In addition, a logo can help distinguish a business from other competitors, explain what the business does, and communicate the overall value of the goods or services offered. In the next two subsections, we'll explore the elements of a logo and the different main types.

Exploring the elements of a logo

Now that we understand what a logo is, let's dive further into the elements that make up a logo.

Logos consist of the following elements:

- *Typography*: A great number of logos contain some form of typographic element such as a monogram abbreviation or the full name of the business.

- *Color*: Logos can range from black and white or monochromatic to a multicolor palette capturing the mood or emotion of the brand. Reviewing what we covered on color theory in the *The Emotions of Color* section of *Chapter 6, Color Theory and Application,* can help you master this.

- *Imagery/shape*: The typographic elements of logos are sometimes accompanied by symbols or icons. The shapes that make up the symbol or icon can help support the mood or voice of your brand, similar to the use of color. For example, curves and circles can suggest friendly and happy emotions, while sharp and pointy shapes can connote strength, aggression, or sometimes danger (depending on how and what the shapes are used for).

- *Context*: There are times when a logo is defined with design elements that match the title of the logo in a literal sense. Different elements in the logo can directly relate to the brand or message it represents. For example, if a company specializes in solar energy products, incorporating elements like the sun or a yellow color palette can help convey the message clearly.

Now that we understand what a logo is and the elements that make up a logo, let's move on to the next section on the different types of logos.

Understanding the seven types of logos

In essence, we understand now that a logo is a symbol consisting of text and images. We can create different variations and combinations of the logo with different styles of typography and images. There are roughly seven types of logos. Each one will give your brand a unique vibe and look, so make sure to think it through carefully before you go through all the trouble of creating one.

The primary different types of logos are as follows:

- *Monogram logos* are composed of letters or initials that represent a brand in a simple, easy-to-remember manner. This is great for businesses with long names and words. Some examples include cable networks and TV apps. This is an effective type of logo that's easy to remember and can be used to brand various marketing materials and campaigns.

Here is an example of a monogram logo I created a few years ago:

Figure 14.1 – Monogram logo

- *Wordmark logos* are similar to monograms in that they are font- and type-focused designs. Some examples that come to mind are the logos for popular soft drinks and Google's logo. This is another very effective logotype that's easy to remember and apply for brands in various marketing materials and campaigns.

- *Pictorial mark logos* (*brand marks*) are icons or graphic-based images that can be identified instantly. They usually work best with well-established and trusted companies. You have to consider the meaning and or emotion that you want your *pictorial mark* to communicate to your audience. Some examples of well-known pictorial marks are Apple, Chase Bank, and Shell.

- *Abstract logo marks* are similar to pictorial marks. Instead of representational images, abstract marks use non-representational mark-making that focuses more on mood, expression, shapes, and patterns. Think of logos such as Nike or Reebok. This logo style can give brands a unique meaning and style of expression.

Here is an example of an abstract logo I created years ago:

Figure 14.2 – Abstract logo mark

- *Mascot logos* are stylized drawings and illustrations of characters that act as the face of your company or brand. They can engage customers and create an emotional connection. Examples include the Quaker Oats pilgrim mascot, Captain Crunch, and Tony the Tiger. It may be tough for mascots to act as a stand-alone logo. For example, it may be hard to print a mascot on small marketing and packaging materials if the mascot has lots of fine detail.

The following is a self-portrait of myself as a mascot logo I created years ago:

Figure 14.3 – Mascot Logo

- *Combination mark logos* consist of wordmarks combined with either a pictorial mark, abstract mark, or mascot. It's the most versatile type of logo to advance your brand. It has the potential to become recognized solely by the image if it becomes a lasting trusted brand. Some examples include Wendy's and the Monopoly game (which uses a font and mascot combination).

The following is a combination logo I created to identify my brand. The figure silhouette holding the paintbrush between the two letters is me:

Figure 14.4 – Combination logo mark

- *Emblem logos* typically feature a font enclosed within a symbol or icon that evokes a sense of heritage and tradition. Some examples are badges, crests, and seals. They can be very detailed and complex, but over the years, companies such as Starbucks have successfully modernized their emblem logos to become both timeless and relevant.

That covers this section on the different types of logos you can create. Now we are ready to move on to the next section that details the process of creating a rough idea for a logo.

Sketching out rough ideas and brainstorming

It's a good idea to experiment with some rough ideas and sketches of your logo. At this stage, your creativity tends to be more relaxed and free-flowing. There is also less pressure to finish the logo for a deadline.

> **Important note**
>
> After breaking down the seven types of logos, we can better decide what type to use for each branding and business project we encounter. We also realize the brainstorming process can take varying amounts of time and effort depending on which logo type you need to create. For example, trying to draw and develop a mascot logo can take much longer than creating a monogram logo.

The first stage of brainstorming and sketching out ideas for a logo for your business involves thinking about the type of business you want to establish, the goods and/or services you'll provide, the demographics of your target audience, the name of the business, and the emotions and mood that you want clients and potential customers to remember you for. These are some of the essentials to building any brand, along with the logo that will represent it.

The second stage of brainstorming and sketching ideas involves researching other companies' brands and logos for inspiration. Look up companies offering similar services or goods to your own. Check out websites such as Pinterest or Dribbble.com to get inspiration for your own creations. The more visual references you have, the more potential ideas you'll have to choose from and develop into a logo.

Now that we've covered some of the basics of preparation, let's cover sketching over the next few pages.

The backstory on our imaginary brand More in Common (MIC)

The logo I'm aiming to create is for our imaginary brand *More in Common*. It is a non-profit organization that was assembled to bring different ethnic backgrounds together by providing educational resources, training for careers, and live entertainment events. This is a way that can bring everyone together, and spark the realization that we are more alike, and have More in Common to build better communities with understanding, support, and unity.

Now that we've covered that backstory on our imaginary brand, let's cover sketching over the next few pages.

Sketching out rough drafts

The process is fairly simple.

1. First, I create a **6" x 8"** new document to work on my logo sketches and ideas.

2. Next, I selected the **Wet marker 2** brush included with Photopea.

    Since it has a wide, bold, fluid, and steady tip, it will be good for sketching out some logo ideas, especially if you have wide or large areas that make up the logo lettering elements. See *Figure 14.5*:

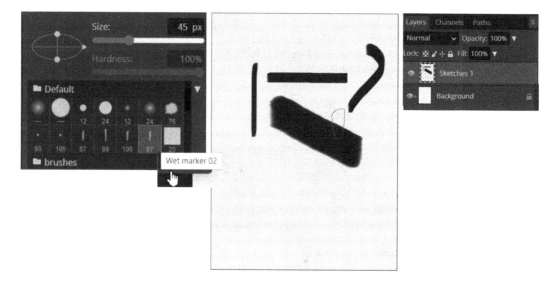

Figure 14.5 – Create a new document and choose Wet marker 02

3. Next, create a *new Layer* and rename it *Sketches 1*.

    Start sketching your ideas out. Be loose and relaxed, and don't worry about messing up. I usually try to come up with 6-10 rough sketches (sometimes more), until I feel I've peaked with ideas. See *Figure 14.6*:

Figure 14.6 – Start sketching out rough ideas

At times you may have a logo you feel is going in the right direction, but you need to create slightly different versions of that particular logo design until you're satisfied with it. I had to do that for the logos in this section.

After making the first batch of rough sketches, I felt the need to try to work out more sketch ideas on a separate page.

Important note

You don't have to do *step 3* unless you feel you can come up with more ideas. As you continue practicing and creating logos, more ideas may come to you later on, at any given moment. Don't stress yourself out if you don't come up with anything you like right away.

4. For the second page, I used the **Pen** Tool and **Shape** Tool to explore shapes that I can incorporate into my logo. I also typed out the **MIC** lettering in a few different fonts that came with Photopea. This helped me build on ideas that I might not have thought of solely from my imagination or memory. You can explore other websites to look at different fonts and logos to help you get creative as well. See *Figure 14.7*:

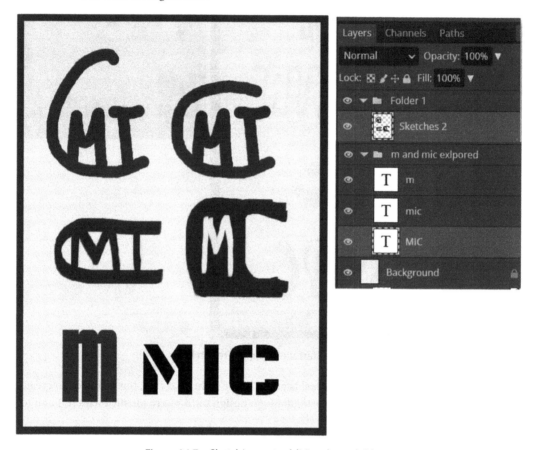

Figure 14.7 – Sketching out additional rough ideas

After reviewing all of the sketches; I decided to polish up the black and white versions of the two rough logo sketches on the right side of *Figure 14.7*. I will create a third logo that's slightly different from the two polished ones in the upcoming section.

Creating a black and white logo

After working on a variety of sketches and rough ideas for the MIC logo, we should narrow down our choices and further develop a few of the logos into polished designs. We will explore a few of the ways we can achieve a polished black and white logo in this section.

Over the years, I've learned it's best to work in black and white so that you don't spend a lot of time (wasted time) on the small details such as how many colors to use, whether you should add gradients and drop shadows, and so on. You can do all of that now, but may not like the design by the time you finish. By leaving all of those details until last, you can focus on the important parts of the design with a clear mind.

With that being said, let's get started on finishing one of the three best logos from the sketches. I have included the example *Figure 14.8* in the resources folder if you would like to follow along.

1. I will use a combination of tracing with the **Pen** tool, using the **Text** tool to type out some of the letters, and adjusting them manually.

2. Select the **Pen** tool to begin tracing the **M**.

3. After tracing the **M** with the **Pen** tool, click **Selection for Pen Mode**.

    I am tracing the **M** in separate sections to make manual adjustments easier when needed. See *Figure 14.8*:

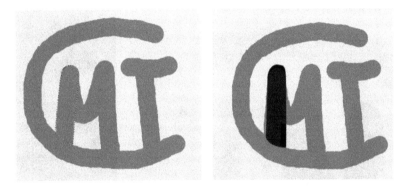

Figure 14.8 – Use the Pen tool to trace the logo in separate sections

4. Next, *duplicate* the left side of the **M** layer and *drag* it to the right side of the canvas.

5. While the **M** layer is active, go to the **Edit** menu and select **Transform | Flip Horizontally**. Next, select the **Move** tool and hold the *Ctrl* key while *dragging* the right side of the **M** down to make it long enough to overlap the **C**. See *Figure 14.9*:

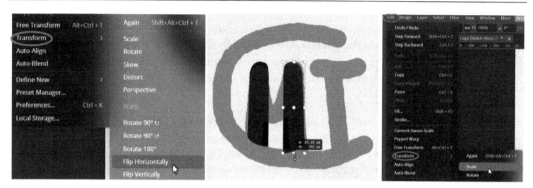

Figure 14.9 – Duplicate the left side of the M, flip it horizontally, and stretch it

6. Now we need to trace the center of the **M** with the **Pen** tool. Make any adjustments to the logo with the Pen tool where needed. For example, I had to make sure the top and bottom of the center of the M had a smooth curve and were even on both sides. See *Figure 14.10*:

Figure 14.10 – Trace the center of the M evenly and smoothly

Let's move on to the letter **I**.

7. Type in **22 px** for the **Corner Radius** value of the rectangle. This will round the corners of the rectangle.

8. Select the **Rectangle** shape tool and create two separate rectangles, one for the top of the **I** and one for the main body of the **I**. (The **C** will create the bottom of the **I** in the next step.)

Early on in my use of Photopea, I used to draw the rectangle first and then tried editing the corners, with no results. Please note that you have to type in the number value *first*, before drawing the rectangle, to make the corners round. See *Figure 14.11*:

Figure 14.11 – Use the Rectangle shape tool to draw the I

Now we can incorporate the **C** into the logo.

9. Select the **Type** tool and type out the letter **C** in a font similar to the style of the **M** and **I**. See *Figure 14.12*:

Figure 14.12 – Use the Type tool to incorporate the letter C into the logo

10. After creating the **MIC** logo, I created another next to it on the right, solely using the **Type** tool. I like the one with the **Type** tool better. See *Figure 14.13*:

Figure 14.13 – The finished logo, and a second logo created solely with the Type tool

Notice how I tried tying all three letters into some interesting abstract shapes, while the other one was just simple text. These logos are *monograms* and *letter marks*.

The next example will be an *abstract logo mark*.

Creating an abstract logo mark

Now it's time to create one more stylized logo to help us decide which one is best suited for our imaginary brand *More in Common* based on the *rough sketch* in *Figure 14.7*, located on the far right of the second row:

1. To get started, *select* the **Type** tool and type out the letter C. *Duplicate* the letter C two or three times. This will enable you to browse through the fonts installed in Photopea and save the best three for the C.

2. I chose the best C for the logo from the three I created and made a *copy* of the **C** text layer.

> **Important note**
> I saved a backup of the letter C just in case I forgot the font or lost the editable text layer. It will also save me time trying to go through all of the fonts again to figure out the name of the font I was using.

3. Next, *right-click* on the **C** text layer and select **Rasterize**, as shown in *Figure 14.14*:

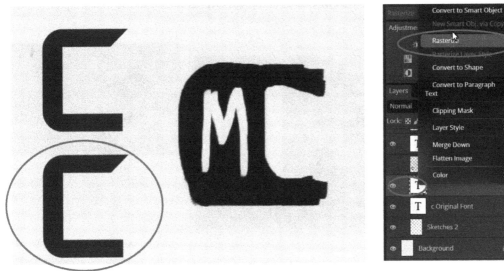

Figure 14.14 – Creating an abstract logo mark

4. Now that we made the **C** a rasterized layer, we will re-purpose it as the base of our logo.

5. Select the **Rectangle Select** tool and drag it vertically over the ends of the **C**, (see *Figure 14.15*).

6. Press the *Delete* key to make both of the ends of the **C** even.

7. Next, go to the **Edit** menu, then **Transform**, then select **Scale**.

8. Hold the *Ctrl* key while dragging the **C** horizontally to stretch it (or widen it vertically to about twice its original size). See the following example. See *Figure 14.15*:

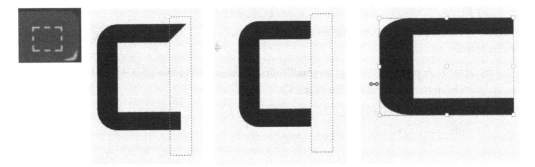

Figure 14.15 – Even the ends of the C and stretch the C horizontally

9. Next, let's fill in half of the opening of the **C** with black. See *Figure 14.16* for an example.

10. Select the **Rectangle Select** tool and make sure the foreground color is *black*.

11. Go to the **Edit** menu, choose **Fill Color**, and press **OK**, as shown in the following example. See *Figure 14.16*:

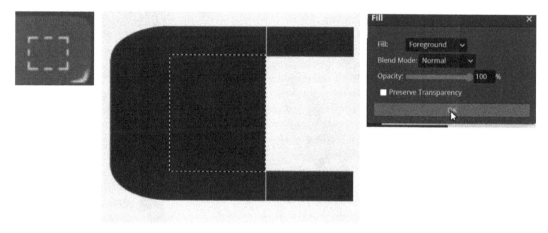

Figure 14.16 – Select the Rectangle Select tool and fill in half of the opening with black

Well done! Next, we need to round off the *corners* of the **C**.

Saving, editing, and exporting the logo into Affinity Designer

Typing out words, or in this case a letter *C*, can be a good starting point for your logo, yet it still may not be quite the look that you want. In this case, there were parts of the *C* I liked, but I needed to customize it more to what I wanted. Follow along to see how and why I did it in the following steps:

1. Select the **Rectangle** shape tool and enter **22 px** for the **Corner Radius** value (to round the edges of our rectangle).

2. Draw a rectangle shape just wide enough to cover the end of the **C** by dragging the mouse to the right.

3. Next, refer to *Figure 14.17*; *drag* the small round rectangle over the edge of the **C** until it lines up (matching the height and edge of the **C**).

4. Duplicate the round rectangle and drag it down to the lower end of the **C**, mirroring the position of the upper one. See *Figure 14.17*:

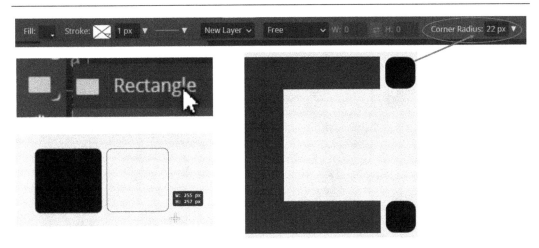

Figure 14.17 – Create a round rectangle and duplicate it to match the corners of the C

5. Now select the **C** Layer to make it the *active* Layer.

6. Select the **Rectangle Select** tool, drag it vertically over the corners of the **C**, and press the *Delete* key.

7. By deleting the square corners on the **C** character, our rounded rectangle selections rounded the ends of the **C**. See *Figure 14.18*:

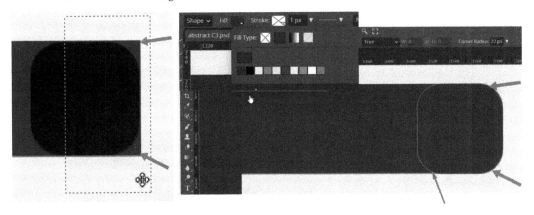

Figure 14.18 – Delete the original square corners of the C and place the rounded squares there

8. Next, *select* both of the small round *Rectangles*, along with the **C**. Then, *right-click* and select **Merge Layers** to merge them into one shape, as shown in the following example. See *Figure 14.19*:

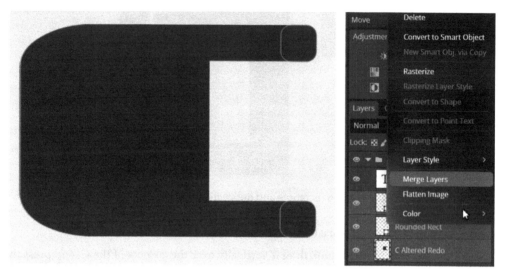

Figure 14.19 – Select the round rectangles and the C to merge them into one shape

9. Now that we've created the shape of the **C** character, we can work on the letter **M**.

10. Select the **Type** tool and type out the letter **M**.

11. *Duplicate* the **M** several times, and pick out and apply the best three fonts that you think work well with the **C** shape we created.

I chose my favorite font from the selection for the **M** and made the letter the right size to fit inside the color-filled area of the **C**, as shown in the following example. See *Figure 14.20*:

Figure 14.20 – Use the Type tool and choose an appropriate font for the letter M

12. After finding the right font for the **M** I double-clicked on the **Type** tool and chose **Color Fill**, located at the top near the main menu bar.

13. To choose the color for the color fill, you can select the **Eyedropper** tool to sample the color of the background, or choose whichever color you prefer for the text color for the **M**.

After placing the **M** inside of the C shape, we can see the letter **I** in an abstract form, located in the white space between the **M** and the **C**, making **MIC**. Have a look at the following example to see what I mean: See *Figure 14.21*:

Figure 14.21 – Change the color of the M with the Type tool's Color Fill option

Now that we have the third *MIC* logo created, I feel this is the logo that should serve as the logo for our brand.

14. Next, we can convert it into a *vector* graphic format. This will be useful if or when we need to get it printed for large-format purposes such as signs, billboards, car decals, and apparel.

15. Go to the **Image** menu located at the top, select **Vectorize Bitmap**, and click **OK**, as shown in the following example. See *Figure 14.22*:

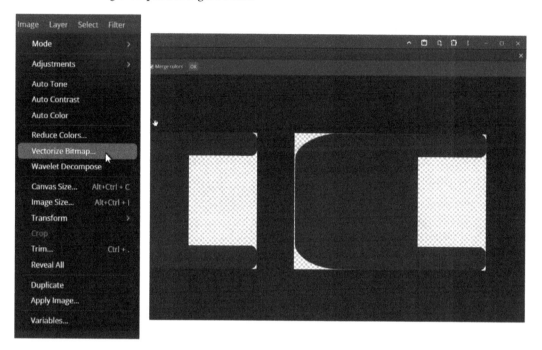

Figure 14.22 – Convert the raster C shape into a vector graphic with the Vectorize Bitmap option

16. Now that we've made the C shape a vector graphic, let's convert the **M** character to a *vector* shape. *Right-click* on the **M** text layer and select **Convert to Shape**, as shown in the following example. See *Figure 14.23*:

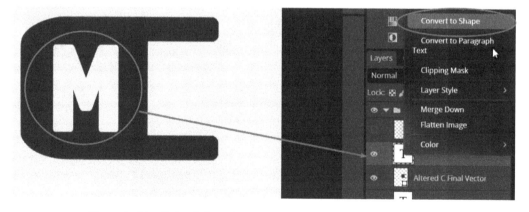

Figure 14.23 – Right-click on the M text layer and click Convert to Shape

17. After taking another look at the *MIC* logo, I realized it would look better if I *decreased* the *height* of the logo and added the full name of the company, *More in Common*.

18. To edit the height of the logo, go to the **Edit** menu and choose **Transform | Scale**.

19. Press and hold the *Ctrl* key while *dragging* the logo down to *decrease* the height of the *MIC* logo. See the following example for the results. See *Figure 14.24*:

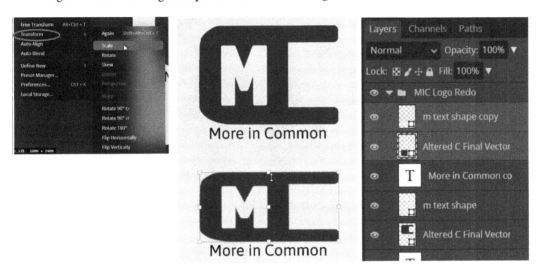

Figure 14.24 – Decrease the height of the logo and add the full business name below

20. Now that the *MIC* logo is finished, we can export the vectorized bitmap logo as an SVG or PDF, or as a vector file that can be opened in other vector-based applications and software such as Adobe Illustrator, Corel Draw, Affinity Photo, and Inkscape, among others.

21. Go to **File** | **Export as** | **PDF** | **Save**, as shown in the following example. See *Figure 14.25*:

Figure 14.25 – Export the vectorized logo as a PDF

After exporting the MIC logo as a PDF, I opened it in Affinity Design, a vector-based program, to see if the file would open properly.

I was able to open the logo and also added some additional tweaks.

> **Important note**
>
> I could have done this in Photopea, but it was much easier to make these additional adjustments in Affinity Designer. Affinity Designer's tools are suited specifically for this, working much more intuitively and smoothly. These tools are my preference when designing logos.

Editing the logo in Affinity Design

For the fine-tuning tweaks, I adjusted the anchor points along the edge of the C to make the shape smoother, deleting some of the anchor points and adjusting a couple of the curves, as shown in the following example. See *Figure 14.26*:

Figure 14.26 – Open the exported PDF file in your preferred vector application

I took one more look at the logo and increased the size of the **More in Common** text to better fit the width of the MIC shape. See *Figure 14.27*:

Figure 14.27 – Increase the size of More in Common to match the MIC shape

Now that we've edited the logo more to our liking, let's move on to the next section.

Exploring colors and fonts

After editing the size of the *More in Common* logo, I decided to try out a variety of color schemes that would work nicely for the logo and brand. But how can we narrow down the appropriate fonts and colors for our final logo? Let's see in the following sections.

Choosing the colors

Certain colors can convey specific meanings. Here are a few of them:

- The color *black* can give viewers and potential clients a sense of mystery, power, class, and strong emotion

- The color *green* can create a sense of nature, calmness, healing, and hope

- The color *gold* gives a sense of warmth, success, and confidence

- The color *white* can create a sense of cleanliness, purification, and hope, and encourages mental clarity

Next, let's choose the right font.

Choosing the font

I chose a *sans serif* font to communicate to viewers who like to challenge themselves; that is, people who don't mind going beyond their traditions and comfort zones. Due to having an open mind and always looking to learn something new, they love to be around people of different backgrounds, nationalities, regions, beliefs, and customs because they realize we have **More in Common**. We are here to experience and enjoy life. The more we realize that as a whole, we can build better communities built on class, calmness, healing, hope, warmth, success, and confidence (as mentioned in some of the emotions colors can engender).

Let's experiment with some of the color schemes with the MIC logo.

The color schemes were executed effortlessly in Affinity Designer. It was best to create the color options in the vector format to avoid any issues with poor resolution if we needed to enlarge them for any printing or large-scale advertisements. See the following figure for the results. See *Figure 14.28*:

Figure 14.28 – Some color schemes created in Affinity Designer

Finally, let's see how the final logo will show up on merchandise.

Creating some MIC logo mockups

Another way to test the logo is to see how it looks as a mock-up. I chose a black baseball hat and jacket to test it. See *Figure 14.29*:

Figure 14.29 – Displaying the MIC logo on some merch mockups

I think the logo looks nice on the jacket and hat; it gives a sense of comfort, friendliness, and respect. That wraps it up for this section. Now let's sum up everything we covered next.

Summary

In this chapter, we covered a lot on how to create a logo. There is a lot more to learn about logos than can be taught in one chapter, but you will have a much better understanding of what's involved after all we've covered in this chapter.

We learned what a logo is, and broke down the elements of a logo, including typography, color, and imagery. We delved even further into logos by looking at the seven types of logos. Understanding the different types of logos can help us make the best choice for which type of logo will be best for a particular brand.

After gaining an understanding of logos, we created a logo step by step, from sketching out rough ideas and brainstorming, to creating a polished black and white logo, and then adding colors and choosing fonts.

If you want to learn more about logos, I recommend taking a course or workshop. I also recommend the book, *Logo: The Reference Guide to Symbols and Logotypes*, by Michael Evamy. It has well over 400 pages of some of the greatest logotypes ever designed. You'll definitely learn a lot and be inspired just by looking through this book.

In the next and final chapter, *Tips, Tricks, and Best Practices*, we will further enhance our skills, knowledge, and relevance when working in Photopea and on image editing more broadly.

Tips, Tricks, and Best Practices

In this chapter, we will explore smart objects and other features of Photopea that you may not be aware of, such as working with the vanishing point filter features, layer comps, and building a portfolio for a career.

By the end of this chapter, you'll be able to get started with adding basic animation to images and drawings. You'll have an understanding of how to use the vanishing point filter, and know how to use smart objects effectively in your documents. In addition, you'll know how to create layer comps to show different variations and arrangements of illustrations that can be applied to other projects such as logos, page layouts, and so on. Lastly, you'll gain insight into tailoring portfolios for the type of work you are interested in and know the different ways to set up and arrange portfolios for print and the web.

In this chapter, we will cover the following topics:

- The vanishing point filter
- Using layer comps
- Exporting layers
- Understanding and using smart objects
- Creating a portfolio for your career

Now that we have an idea of what to expect in this chapter, let's get started with the first section.

The vanishing point filter

One of the most powerful image editing tools in Photopea is the vanishing point filter. This tool allows you to define 3D areas on top of your images, creating a dynamic level of creativity and precision.

Once the 3D areas are defined, you have the flexibility to place content directly into these spaces with precision. Whether it's inserting text, images, or graphic elements, the Vanishing Point filter ensures seamless integration within the perspective of the image.

Let's get a better understanding of this filter by trying it on an example.

Using the Vanishing Point filter

Let's practice using it by following these steps:

1. Grab the photo **Empty Room Wall Photo** and **M Earth VP** file from the `Unlock Your Creativity Resources` folder.

2. To access the **Vanishing Point** filter, go to the **Filter** menu and select **Vanishing Point…**. Once you open the **Vanishing Point** window, you will see your document with the current Layers at this point, but we have to do a few more things for the filter to work.

3. To begin, exit out of the **Vanishing Point** window (if you currently have it open) and return to the default window of the original document.

4. In my example (*Figure 15.1*), you see a photo of an empty bedroom, and the Layer name is **Bedroom**.

5. Duplicate the **Bedroom** Layer as a copy for backup.

6. Next, create a new (empty) Layer titled **Layer 1** above the **Bedroom** Layer, and keep it active (selected). **Layer 1** will be the layer to which you want to apply the **Vanishing Point Filter**.

7. Now, go to the **Filter** menu and select **Vanishing Point…** to launch the **Vanishing Point** window. See *Figure 15.1*:

Figure 15.1 – Creating a new layer and launching the Vanishing Point filter

8. Next, select the **Create Plane** Tool just below the **Move** Tool located at the top left and a red line or angled path to match the planes of the wall will appear. You will need to drag the red line to each corner of the wall until it lines up and prompts a blue grid. See *Figure 15.2*:

Figure 15.2 – Using the Create Plane tool to match the planes on the wall

9. The blue grid means that the wall area is selected correctly in perspective and is ready to place the images on the wall. See *Figure 15.3*:

Figure 15.3 – Creating a new layer and launching the Vanishing Point filter

Now that we know how to set up a **Vanishing Point** grid, let's get started with the next steps.

Adding the first image to the wall

Now, we need to get our images placed on the wall:

1. Press **OK** to return to the main window.

2. Next, open the image you would like to place on the wall from your desired file (or image).

3. Next, go to the **Select** menu and choose **All** (or press *Ctrl + A* for Windows or *Command + A* for Mac users). Then, go to the **Edit** menu and copy the **M Earth VP** image (or your own image). See *Figure 15.4*:

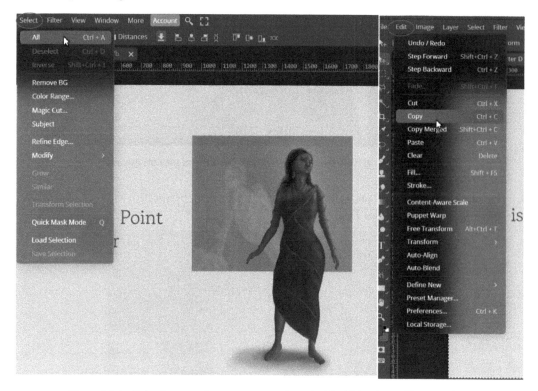

Figure 15.4 – Selecting your images to paste into the Vanishing Point filter window

4. After you've made a copy of the images, return to the initial document with the **Bedroom** Layer, select the empty **Layer 1**, and go to **Filter | Vanishing Point…** to launch the **Vanishing Point** grid we set up earlier. See *Figure 15.5*:

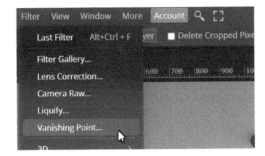
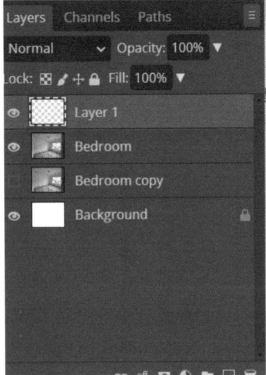

Figure 15.5 – Switching to the main document and pasting the images into the Vanishing Point window

5. Next, paste the **M Earth VP** Layer (image) into the **Vanishing Point** window.

6. Notice how the image fits into the same perspective as the wall.

7. While it's selected, drag the corner anchor point down to the corner of the wall, and enlarge the image to fit on the wall in perspective.

8. Press **OK** to return to the main window. Notice that the image (**M Earth VP**) is now on the Layer titled **Layer 1**. You can rename it or leave it as it is. See *Figure 15.6*:

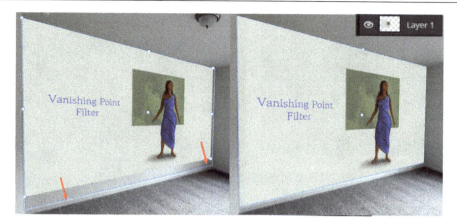

Figure 15.6 – Adjusting the size of the image to fit into perspective

> **Important note**
>
> In *Figure 15.7*, I removed the blue **Vanishing Point Filter** text in the illustration to add something more interesting and relative to the image.

Adding a second image to the wall

Now, we will add a second image to the same wall:

1. To do this, create a new Layer; you can leave it as the default name, **Layer 2** (optional).

2. Next, go to **Filter** and select **Vanishing Point…**. See *Figure 15.7*:

Figure 15.7 – Creating a new layer, Layer 2, and then opening the Vanishing Point window

3. Then, select the next image you would like to place on the wall since there is a lot of empty space.

4. I opened another image of the woman standing, titled **M Earth Greyscale**. It is a portrait that I want to place where the blue text was.

5. With the **M Earth Greyscale** image open, go to the **Select** menu, choose **All** (or press *Ctrl/Command + A*), and click **Copy**, or press the shortcut keys *Ctrl/Command + C*. See *Figure 15.8*:

Figure 15.8 – Opening the second image, M Earth Greyscale, to Select All and Copy

6. Next, paste the image (in my case, **M Earth Greyscale**) into the **Vanishing Point** filter window.

    Notice how the image also fits into the same perspective as the wall but is a bit too big. See *Figure 15.9*:

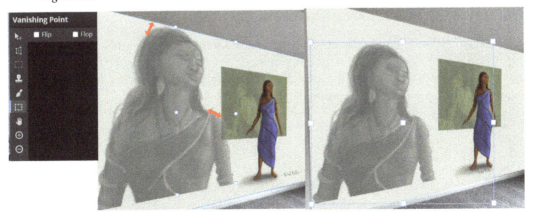

Figure 15.9 – Paste the M Earth Greyscale image into the Vanishing Point window and resize it

7.   While it's selected, drag the corner anchor point down to reduce the image size enough so that it looks balanced next to the full figure on the right (the **M Earth VP** layer).

> **Important note**
> You can still adjust the size of the images and move them around in the default (main) window if you select the **Move** tool.

8.   Click **OK** to return to the main window. Notice that **M Earth Greyscale** is now on the layer titled **Layer 2**. You can rename it or leave it as it is. See *Figure 15.10*:

Figure 15.10 –M Earth Greyscale layer renamed after closing the vanishing point filter

Now that we have both of the images in perspective with the **Vanishing Point** filter, I figured I would go ahead and add my personal signature on the wall using the same.

Adding a personal signature on the wall

Now that I have placed the artwork on the wall in perspective, I decided I want to add the finishing touch with my signature. In a realistic scenario, you would want to brand your work so that people know who created the art. My signature will not be included in the `Unlock Your Creativity Resources` folder version. This is how I did it:

1. To get started, I created a new layer titled **Layer 3**. See *Figure 15.11*:

Figure 15.11 – Creating a new layer, Layer 3, to add my signature

2. With **Layer 3** selected, I went to **Filter** and clicked **Vanishing Point…**.

3. I opened up the **M Earth Greyscale** portrait file to select my signature.

4. I then went to **Edit** | **Select** | **All** | **Copy**.

5. Next, I went to **Edit** | **Paste** (or you can press *Ctrl/Command + V*) to paste my signature on the **Vanishing Point** window.

6.   I resized my signature by going to **Edit | Transform | Scale**. See *Figure 15.12*:

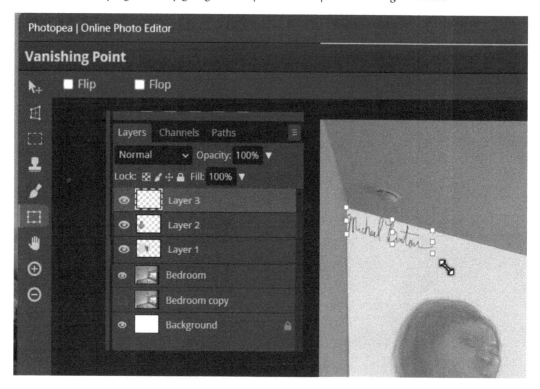

Figure 15.12 – Resizing my signature

7.   After resizing my signature, I moved it to the lower right with the **Move** tool.

We have now finished decorating the wall with the **Vanishing Point** filter. See *Figure 15.13*:

Figure 15.13 – Moving the signature to the lower left of the wall to finish

We can decorate all the wall space if we want to. It would be nice to see what we could come up with for our rooms if we ever decided to paint a mural or get images printed out for wallpaper.

That sums up this section on using the **Vanishing Point** filter. Let's move on to the next section.

Using layer comps

Using layer comps in Photopea is simple, yet it can serve as an incredibly powerful tool that can enhance the way you approach various creative tasks. Whether you are designing website interfaces, creating fashion concepts, developing logos, or working on intricate concept art, layer comps can streamline your workflow and boost your productivity, including cases where you may need to export layers versus the entire document.

A **Layer Comp** is a feature included in Photopea that enables you to save and manage different combinations of layers in your document. Each layer comp stores the visibility, position, and appearance of the layers at a specific moment, position, or layout, making it easier to switch between different versions of a design you can use to present to clients, or projects where you need to make a decision on which works best for the brand or role.

Let's look at how to use cases for layer comps:

- **Designing Website Interfaces**: When designing website interfaces, having the ability to quickly switch between different design variations is crucial. With layer comps, you can easily showcase different color schemes, layout options, and graphic elements without having to make the best design decisions when accompanied by visual comparisons.

- **Creating Concept Art**: For concept artists, layer comps offer a convenient way to explore different versions of characters, creatures, vehicles, or weapons. By organizing your layers into different comps, you can switch between variations, experiment with different design elements, and refine your concepts with ease. This flexibility allows you to iterate quickly and efficiently, leading to a more finished and refined final artwork.

- **Innovating Fashion Design**: In the world of fashion design, where creativity and originality play a major role, layer comps can be a game-changer. By creating comps for different color palettes, fabric textures, and garment styles, you can visualize your designs in various combinations without starting from scratch every time. This not only accelerates the design process but also encourages experimentation and innovation in your creative work.

Now that we have a better understanding of what layer comps are, let's learn how to put them into practice.

Creating Layer Comps

Before setting up the first layer comp, it's important to understand the three key flags associated with each comp: **Visibility**, **Position**, and **Appearance**.

These three flags help manage the different versions of your document. They allow you to alter and switch layers around. You can decide on what element or layer within the document is visible or positioned or whether it appears in a different color, size, filter, and so on, for the best options.

Next, we will use the **Time to Shine** image with six different concepts.

Three of the six concepts will be a step-by-step process that we can create and save as Layer Comps. You can open up your own file, or just follow along with my example.

To create a layer comp in Photopea, follow these steps:

1. Open an image or design that has several layers; I will open my illustration example, **Time to Shine**.

2. Next, go to the **Window** menu and select **Layer Comps** to launch the panel.

3. You will see the **Last Document State** checked at the top of the **Layer Comps** panel.

> **Important note**
> **Last Document State** means that this is the latest state or stage in which the document was altered and saved.

4. You will see four buttons located at the bottom right of the **Layer Comps** panel: **Files**, **New Update**, **New** (to create a new Layer Comp), and a **Trash** icon to delete a Layer Comp.

5. Click the **New** button to create a default state of the current document, give it the name **Original Illustration**, and save it. See *Figure 15.14*:

Figure 15.14 – Creating a new Layer Comp with the current document arrangement

6. Next, we will make some changes to the illustration by hiding some of the layers and folders in the original document arrangement and revealing some of the other layers that have different data or design elements.

Important note

Each time you make an edit with the current state of the document and Layer Comp, I would suggest constantly saving the document, and also pressing the **Update** button that has the two curving arrows and is located at the bottom of the **Layer Comp** panel. This will reduce the confusion of working on the wrong Layer Comp as you continue experimenting with new arrangements.

7. Next, click **New** at the bottom of the **Layer Comp** panel to create a second version of the illustration and rename it **Illustration Version 2**. Make sure this Layer Comp Layer stays active while making changes so the changes apply to this version only.

8. Hide the **Grass** and **Lt Blue Hiji Scarf** Layers, and reveal the **Teal Hiji Scarf** layer. Hide the **Blue Dress** layer and reveal the **Dark Blue Dress** layer. We will close her eyes by hiding the **Shanel Face Smile** folder and revealing the **Eyes Closed** layer.

9. Press the **Update** button again, just to confirm that all the new changes and arrangements are finished for **Illustration Version 2**. You will see a checkmark appear on the **Illustration Version 2** layer as a confirmation that the data is stored and saved.

10. Next, go to **File | Export as** to save the Layer Comp as JPEG, PNG, or whichever format you prefer. I renamed the PNG Time to Shine 2. See *Figure 15.15*:

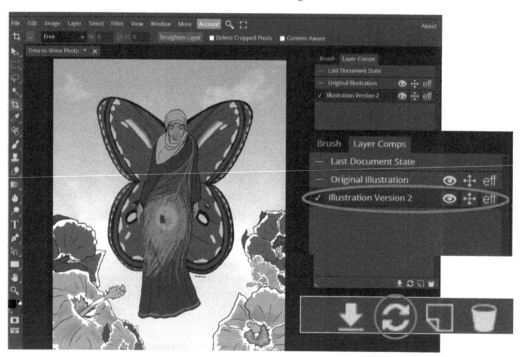

Figure 15.15 – Creating a new arrangement for the second Layer Comp

> **Important note**
>
> You can click the dash mark next to the **Original Illustration** Layer Comp to make sure the original version is still saved.

Now, we are ready to make our third Layer Comp:

1. Click **New** at the bottom of the **Layer Comp** panel and rename it **New Comp 3**.

2. Hide the **Hiji Scarf** layer and reveal the **Long Wavy Hair** layer.

3. Next, select the **Move** tool, and move her toward the top of the document.

4. Press the **Update** button to confirm and then save the data by pressing *Ctrl + S* on the keyboard to save the document.

5. Next, go to **File | Export as** to save the Layer Comp as JPEG, PNG, or whichever format you prefer. You can also rename it to something that distinguishes it from the previous Layer Comps. I renamed it **Time to Shine**. See *Figure 15.16*:

Figure 15.16 – Creating a new arrangement for the third Layer Comp

Now that we've gotten through the three different layer comps, I can show you some other versions I played around with. See *Figure 15.17*:

Figure 15.17 – Creating three additional Layer Comps

After experimenting with the layer comps, you can see how powerful and effective it will be when incorporated into your workflow. It will save you so much time, file size, and space.

Imagine trying to rearrange a design with a lot of layers, effects, and other renditions you would like to experiment with without layer comps. You would have to save each file separately, remember what arrangements you made, open each version separately just to see what each one looked like, and so on.

Overall, layer comps in Photopea are versatile tools that can benefit creatives across various industries. Whether you are a web designer, concept artist, fashion designer, or logo creator, incorporating layer comps into your workflow can enhance your creativity, efficiency, and overall design process.

We are now ready for the next section.

Exporting layers

Now that we have seen how to export the three different layer comps from the same document, let's look at how we can export an individual layer from a document.

Exporting an individual layer from a document allows you to use a single layer for a number of use cases, depending on what you need it for.

For example, you may be working on a team project for a comic book. Let's say that book is about women on the rise. The team member in charge of designing the book cover and interior design may need some elements from the illustrations to design the book cover, and so on. You can export individual layers – for example, **Butterfly Wings** and **Flowers** – separately as PNG files to add around the title of the front cover and parts of the interior pages.

There are several ways to export an individual layer or layers. Let's see how it's done in the next section.

Hiding all other layers

One method to export a layer is to hide all of the Layers you don't need and export the visible Layer or Layers as a PNG. See *Figure 15.18*:

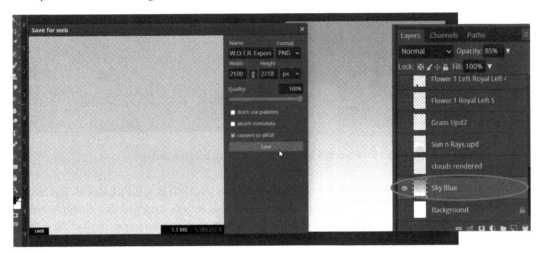

Figure 15.18 – Hiding all layers except the one you want to export as a PNG

Now that we understand how to export a layer with the first method, let's try the second method.

Exporting layers from a Layers folder

The second method to export layers is as follows:

1. Create a new folder and rename it -e-.

2. Create as many folders titled -e- with the number of layers you would like to export.

3. I will place the **Flower Individual** layer in one -e- folder, the **Large Feather** layer in a separate folder titled -e-, and the **Sky-Blue** layer in a third folder titled -e-.

4. Next, go to **Export Layers**, and the **Export** panel will open.

5. You can leave all three boxes checked in blue, under the ---**3 exportable layers** option. See *Figure 15.19*:

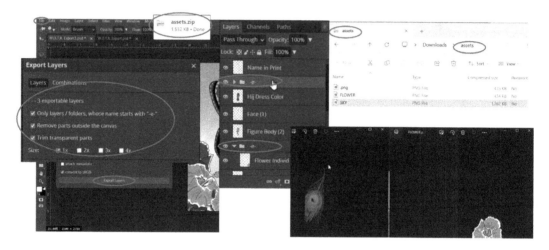

Figure 15.19 – Placing a layer to export into a folder titled -e-

6. You can also *uncheck* the options on the **Export Layers** panel so that it includes all layers in the document, along with the full object on the layer that may be cropped. For example, this one has **17 exportable layers** and includes the full flowers that may be cropped off at the edge of the page. See *Figure 15.20*:

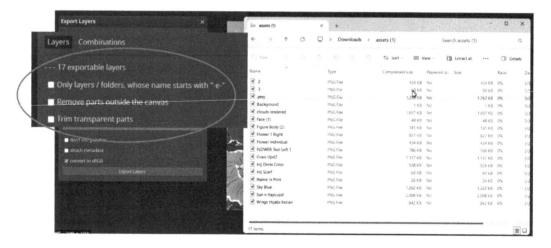

Figure 15.20 – Unchecking all three boxes to export all the layers in the document

> **Important note**
> The more layers you have, the longer it will for the **Export Layers** process to finish.

Let's move on to the third method.

Exporting as a Smart Object

The third method to export a layer is as follows:

1. Right-click on the **Large Feather** Layer and click **Convert to Smart Object**.
2. Next, double-click the **Smart Object** Layer to open it as an individual Layer in a separate window.
3. Now go to **File | Export as | PNG**. See *Figure 15.21*:

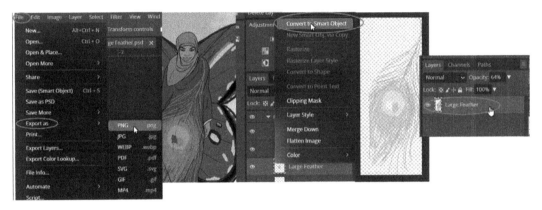

Figure 15.21 – Converting the layer to a Smart Object and exporting it as a PNG

Now that we have covered exporting layers, let's move on to the next section

Understanding and using smart objects

Do you have a design or drawing that constantly uses a specific design element throughout a series of some, or all, of your products or publications?

For example, do you place your logo on all of your branding and promotional materials, such as business cards or a billboard? Or are you a comic book artist who may use an element such as rain, or props that you repeatedly use for your characters? You would benefit from converting them into Smart Objects.

What are Smart Objects?

Smart Objects are a powerful feature offered in Photopea. They allow users to work with image data in a non-destructive manner, retaining all the source content and properties. They can contain raster or vector images and allow users to apply transformations, filters, and adjustments without permanently altering the original image.

The following are some of the benefits of smart objects:

- **Non-destructive Editing**: Using smart objects enables you to make changes to your designs without loss of quality. This can be useful when working on complex projects that require multiple adjustments.

- **Flexible Transformations**: smart objects enable you to scale, rotate, and warp images without the loss of pixel data. This makes experimenting with different compositions and layouts easier and more efficient.

- **Efficient Workflow**: smart objects can streamline the design process with ease when working with linked files. As a result, any changes made to the original file will update automatically in the smart object. This can save you countless hours of time and effort.

Now that we understand the benefits of smart objects, let's move on to the next section.

Creating a Smart Object

For this section, we will create a Smart Object step by step. I will be using some of the flower props from my *Time to Shine* series to demonstrate it.

When creating smart objects, especially from raster images and objects, it may be best to start them on a high-resolution document and **300 DPI**.

Let's create a smart object in the following steps:

1. I drew and painted the flowers on a **7" x 6"** document, at **300 DPI**. This ensures that the flowers won't be pixelated when we bring them into another document unless we greatly reduce the size of the **Hair Flowers** Layer before converting it to a smart object.

2. To demonstrate this, I made a copy of the **Hair Flowers Flat** Layer and reduced it to a very small size, titled **Hair Flowers Flat Sm**. This is located on the left side of the larger **Hair Flowers Flat** Layer on the right side of the document. See *Figure 15.22*:

Figure 15.22 – Making a copy of Hair Flowers Flat and scaling it down to a smaller size

3. Next, select the **Hair Flowers Flat Sm** Layer. Go to the **Edit | Transform | Scale** to enlarge the **Hair Flowers Flat Sm** Layer close to the original size of the **Hair Flowers Flat** Layer. You will notice heavy pixelation after the **Hair Flowers Flat Sm** layer is enlarged. See *Figure 15.23*:

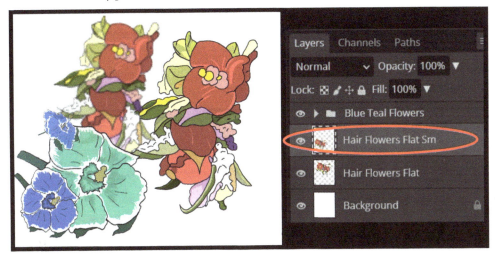

Figure 15.23 – Heavy pixelation after enlarging Hair Flowers Sm

4. After seeing the results of enlarging the **Hair Flowers Flat Sm** layer with heavy pixelation, let's see what happens when we convert the **Hair Flowers Flat** layer into a Smart Object.

5. Select the **Hair Flowers Flat** layer and duplicate it as a backup Layer.

6. Next, select the **Hair Flat** layer, right-click the mouse, and select **Convert to Smart Object**. You should see a small black square on the **Hair Flowers Flat** layer.

7. After making the copy, it's now titled **Hair Flowers Flat copy 2**; it is also a Smart Object.

8. Next, greatly reduce the size of **Hair Flowers Flat copy 2**. See *Figure 15.24*:

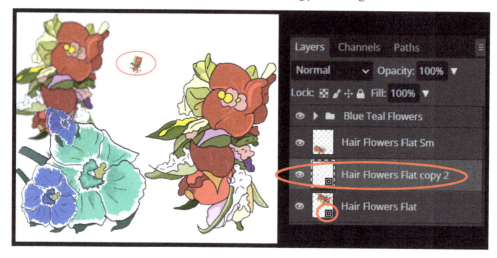

Figure 15.24 – Converting Hair Flowers Flat copy 2 to a Smart Object and reducing its size

9. Now, enlarge the size of **Hair Flowers Flat copy 2** close to the same size as the original **Hair Flowers Flat**.

10. You will see that it retained its high-quality look, due to the smart object retaining all of the data on the layer. See *Figure 15.25*:

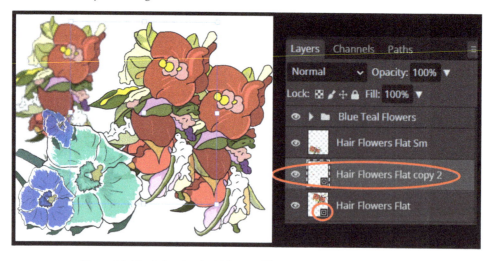

Figure 15.25 – Enlarging Hair Flowers Flat copy 2 retained its quality

11. I went ahead and made a copy of the **Blue Teal Flowers** Layer, and converted it to a smart object as well. Since each flower is separate inside the folder, we can create independent smart objects of each flower if we choose to later. See *Figure 15.26*:

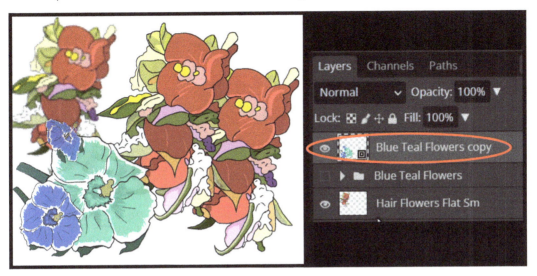

Figure 15.26 – Converting Blue Teal Flowers copy into a Smart Object

Now that we understand smart objects, let's take them a step further and see how we can use them for other situations in the next section.

Additional things we can do with Smart Objects

In addition to smart objects being non-destructive, preserving the high quality of the layers, you can apply filters, effects, and so on non-destructively.

For fun, let's bring the flower props into the illustration of the woman that we used in the *Using the Vanishing Point filter* section of this chapter:

1. Selected both the **Blue Teal Flowers copy** and **Hair Flowers Flat** Layers and press the shortcut keys *Ctrl + C* to copy them.

2. Next, open up the illustration we used for the vanishing point filter exercise of this chapter.

3. Press *Ctrl + V* to paste the flowers into the document.

4. You will notice there are small splats of paint to the left of the **Hair Flowers Flat** Layer. We cannot remove the splats of paint while the flowers are a Smart Object.

5. To resolve this, right-click on the **Hair Flowers Flat** Layer, right-click again, and select **Rasterize**. See *Figure 15.27*:

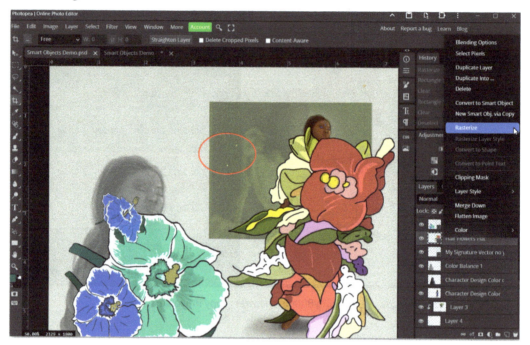

Figure 15.27 – Rasterizing the Hair Flowers Flat layer to remove brush splats

> **Important note**
> When you rasterize a layer, it removes the Smart Object symbol.

6. We can remove the splats of paint with the **Eraser** Tool or use the **Rectangle Select** Tool to make a selection around the area of the splats and press **Delete**. See *Figure 15.28*:

Figure 15.28 – Using the Rectangle Select tool to select and delete any paint splats

7. Afterward, select the **Hair Flowers Flat** Layer, right-click on the Layer, and select **Convert to Smart Object**. If needed, we can rasterize the **Hair Flowers Flat** Layer again at any time. See *Figure 15.29*:

Figure 15.29 – Converting the Hair Flowers Flat layer back to a Smart Object

8. Next, reduce the size of the **Hair Flowers Flat** Layer.

9.   To do this, go to **Edit** | **Transform** | **Scale** to reduce the size small enough to place the flowers in the hair and on the wrist of the character titled **Character Design Color**. See *Figure 15.30*:

Figure 15.30 – Reducing the size of the flowers to fit on the character's hair and wrist

One thing we can probably do is make the flowers hard-edged (the **Hair Flowers Flat** flowers look more painterly and simplified) by applying some filters and adjustments to the **Smart Object** layer.

10.   Go to **Filter** | **Stylize**, and select **Oil Paint**. We can make adjustments to the **Oil Paint** filter by moving the sliders for **Radius**, **Cleanliness**, **Scale**, and **Bristle Detail** until we feel that it complements the painterly style of the character. See *Figure 15.31*:

Figure 15.31 – Adding the Oil Paint filter to the Hair Flowers Flat layer

11. In addition to applying the **Oil Paint** filter, we can add the **Dust & Scratches…** filter to **Hair Flowers Flat**.

12. Go to **Filter | Noise | Dust & Scratches…**.

13. Next, adjust the sliders for **Radius** and **Threshold** to further stylize and soften the **Hair Flowers Flat** layer. See *Figure 15.32*:

Figure 15.32 – Adding Dust & Scratches to Hair Flowers Flat

14. After applying the **Dust & Scratches** filter, we can try the **Gaussian Blur** filter as a final touch to the **Hair Flowers Flat** layer.

15. First, make a selection around the hair flowers by pressing and holding the *Ctrl* key and clicking on the **Hair Flowers Flat** Layer to activate the selection. See *Figure 15.33*:

Figure 15.33 – Making a selection around the Hair Flowers Flat layer

16. Next, go to **Filter | Blur | Gaussian Blur**.

17. Adjust the **Radius** slider until you are satisfied with the soft edges and painterly look. See *Figure 15.34*:

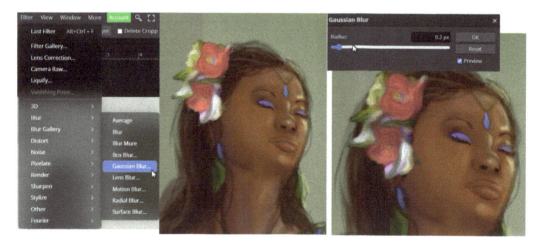

Figure 15.34 – Applying the Gaussian Blur filter to the Hair Flowers Flat layer

18. Repeat the process for the flowers on the wrist. (We may not keep them for the final illustration.) See *Figure 15.35*:

Figure 15.35 – Repeating the entire process of adding the filters to the wrist flowers

19. We can enhance the character design illustration by making copies of the **Hair Flowers Flat** Smart Object Layer and adding them to the background.

20. After making copies of the flowers, we can create a **Black & White** adjustment layer to make them black and white. See *Figure 15.36*:

Figure 15.36 – Creating a Black & White adjustment layer for Hair Flowers Flat copy

21. Next, enlarge, move, and position the flowers we copied as design elements, to make the illustration more interesting. This also shows you another way to use Smart Objects. See *Figure 15.37*:

Figure 15.37– Enlarging the flowers and positioning them as design elements

That sums up this section. We now see the benefits of converting an image or prop to a smart object. This is due to smart objects being able to save and retain larger files of data on the hard drive of the computer rather than storing it in the Photopea document.

Now, we are ready to move on to the next section.

Creating a portfolio for a career

We have covered a lot of material throughout this book. We have learned how to access and use the tools and effects in Photopea. This included things such as navigating the workspace, creating documents, using and managing layers and masks, making selections, using the Brush tools, and so on. We have learned how to make image composites and marketing promos for Instagram. We went from learning the fundamentals of drawing and painting to creating a logo and creating a simple animation in Photopea.

You may be wondering what direction you should take to create a portfolio for a career.

To help give some clarity to your question and this often-discussed topic, let's break it down in the upcoming sections.

What is a portfolio?

In general, a portfolio is a collection of your best work, presented with a visual appeal and well-thought-out layout. It allows potential clients and employers to view your design style, skills, and creativity in a short amount of time. Your portfolio should display a range of projects that demonstrate your versatility and expertise in different areas that align with the creative field you want to work in.

Now, let's look at three essential things to know about portfolios:

- **Portfolio Size**: When creating a portfolio, it's important to consider the intended use of the final product. Portfolios come in various sizes, and the size you choose will depend on whether you plan to display it online or in print.

- **Online Portfolios**: If your portfolio is meant to be viewed online, you have the flexibility to choose any size you prefer. However, it is recommended to save the images at 72 DPI to ensure quick loading times for viewers.

- **Printed Portfolios**: For printed portfolios, it is crucial to select the right dimensions to showcase your work effectively. The most common sizes for printed portfolios are `8.5" x 11"` and `11" x 17"`. When preparing images for print, it is advisable to use a resolution of `100-150 DPI` (going to `200-300 DPI` is optional and, in some instances, better, depending on the type of details in the graphic) to maintain image quality.

Keep in mind that the size of your printed portfolio should align with the type of work you have created and how you plan to present it. Certain projects may benefit from a larger format, while others may be better suited to a smaller size.

Choosing a creative field to build your portfolio

Since we covered a lot of different creative projects in this book, think about which projects you were engaged with the most. Which assignment, or assignments, did you enjoy creating? This will lead you to explore some of the different creative fields you can work in today.

> **Important note**
>
> I will only cover some of the creative fields that I feel may overlap with the exercises and material that we've covered in this book. There is too much involved that cannot be explained fully in a section of a chapter. I will provide some additional resources and links at the end of the section in case you want to learn more about them, as well as types of portfolios.

Let's look further into some of the creative fields, including what they are and what creatives do in particular creative fields.

Graphic design

Graphic design is a discipline that involves composing and arranging visual elements for a wide range of projects. These visual elements consist of typography, imagery, color, and form that are composed to communicate a message hierarchically. There is a wide range of other graphic design disciplines, as well as other creative fields that require different sets of skills, and yet can overlap because one type of graphic design can rely heavily on the other to communicate the message in different media formats.

Let's look into the different areas of graphic design to gain a better understanding of them to aid in building a portfolio of consistent work and also in choosing an area of interest.

Different types of graphic design

Here is a list of some different types of graphic design:

- **Visual Identity Design**: This type of graphic designer works closely with brand stakeholders to create various assets that represent a brand. You may design standard business cards, corporate stationery, logos, typography, color palettes, and image libraries. These assets are set up as visual guidelines called **style guides**, tailored for a company's story and personality, and used throughout different media platforms.

- **Publication Design**: Publication designers create visually appealing layouts for both printed and digital materials for books, newspapers, magazines, reports, catalogs, and so on. They play a crucial role in shaping the way information is consumed by audiences.

- **Motion Graphics**: Motion graphic designers bring static images to life. This dynamic form of visual storytelling encompasses various design elements, such as animation, audio, typography, imagery, video, and other effects. These elements work together to create engaging and captivating content used in animated logos, movie trailers, video games, apps, and so on.

Illustration

Illustration captures the essence of a story from drawings and paintings that are developed into narrated images.

Illustrators are needed for a vast number of things, such as films and games, storyboards (for films, TV series, and commercials), editorials, magazines, posters, fashion and costume design, picture books, and so on.

Different types of illustration

There are a vast number of artistic styles used for illustration. They include the following:

- Commercial art
- Caricature
- Cartoons and comics
- Anime
- Fashion design
- Fantasy art
- Fine art
- Concept art – this uses some aspects of illustration

Commercial art

Commercial art is an area of illustration that crosses over into image editing and graphic design. Initially, it's where creatives develop visuals for advertising agencies and graphic design studios. The illustrator will create either from their personal style of art or they may be versatile to produce a work similar to a style a brand needs.

The design work is created based on a design brief, the target audience, marketing, and branding of the company or client. You can find out more about the different areas of illustration online, and also with a link I provided in the *Resources for creating a portfolio* section.

Let's move on to the next area.

Concept art

Concept art is the initial stage of visual development where artists explore and define the look and feel of characters, environments, or props. It is the idea of what these elements might look like, providing a visual blueprint for the project. Concept art serves as a guide for the design team, helping them to establish a cohesive visual style.

What distinguishes concept art from illustration is that goes beyond just presenting design; it takes those concepts and adds context, emotion, and visual narration to tell a story.

Different types of concept art

Here's a list of different types of concept art:

- Character
- Creature design

- Environment

- Vehicle

- Weapons and props

- Digital matte painting

Digital matte painting

Digital matte painting is a dynamic art form involving the creation of imaginary or realistic sets created by mixing and mashing photos, 3D assets, and digital paintings that are created from a realistic approach to achieve believability. It overlaps with techniques for image compositing. See *Chapter 9, Exploring Advanced Image Compositing Techniques*. Matte painting techniques make it possible for camerapeople and producers to film scenes for genres such as fantasy and science fiction that would be nearly impossible to achieve in real-life environments.

This form of concept art stood out to me because I did something similar to this in *Chapter 9, Exploring Advanced Image Compositing Techniques*, in the *Applying skills to an image composite – man walking*. I mixed and mashed images from several photos, and used the Brush tool to paint in some small areas where needed.

That sums up this area of exploring different creative fields to help decide which path you want to concentrate on to build up a portfolio. Let's move on to the next section, where I answer the question of how many pieces to include in a portfolio.

How many pieces should I include in a portfolio?

If you're a beginner, it's best to learn and work consistently with one specific creative field and minor in another one. After spending time in concept art forums and watching videos from seasoned veterans, most students study intensely for two or more years. This gives you some time to build up an entry-level portfolio of your best 10+ projects.

After you gain some skills and explore a few different subjects, it would probably be best to pick a story or theme where you can build a consistent body of work that can help you reach your goal of 10+ projects.

For example, while exploring concept art, I created a fantasy art story based on conservation and fantasy characters that represented different elements and environments of the planet Earth.

One of the characters is a tree creature with a human anatomical form. His story is to keep the trees in balance to help sustain the planet. I can build a series of illustrations and concept art based on this character in different scenarios to build up a cohesive body of work. After I finish this body of work, I can explore other subject matter to show my range and imagination when creating work for different subjects under concept art.

Now that we have a better understanding of how many pieces to add to your portfolio and how to approach and achieve it, let's move on to the next section.

Resources for creating a portfolio

There are a lot of great online resources to learn more about the different industries and help with creating a portfolio.

I have provided some resources here:

- *How to Create a Portfolio and Get Hired: A Guide for Graphic Designers and Illustrators* by Fig Taylor
- *Creating a Graphic Design Portfolio*: `https://www.format.com/magazine/resources/photography/graphic-design-portfolio-tips`

Here are some portfolio websites:

- *Over 8,000 curated websites on Site Inspire*: `https://www.siteinspire.com/websites?categories=19`
- *5 Designer's Portfolios You Must Checkout in 2024*: `https://medium.com/@arpitchandak18/5-designers-portfolios-you-must-checkout-in-2024-cd72a35deb61`
- *How to Create a Portfolio and Get Hired, A Guide for Graphic Designers and Illustrators*: `https://www.amazon.com/How-Create-Portfolio-Hired-Second-dp-1780672926/dp/1780672926/ref=dp_ob_title_bk`

Learn more about the different graphic design fields here:

- *The 8 Types of Graphic Design You Need to Know*: `https://99designs.com/blog/tips/types-of-graphic-design/`

Learn more about the different illustration fields here:

- *15 types of digital illustration to help you get started*: `https://www.cgspectrum.com/blog/different-types-of-illustration`
- *Graphic Design Trends 2024*: `https://dribbble.com/tags/graphic%20design%20trends%202024`

Learn more about the different concept art fields here:

- *An Intro to Matte Painting*: `https://www.clipstudio.net/how-to-draw/archives/156700`
- *IAMAG Master Classes – Concept Art, Animation, VFX, Storyboard, Career Management*: `https://masterclasses.iamag.co/`
- *Feng Zhu Design School*: `https://fzdschool.com/`

That is all we will cover for this section. There are literally thousands of resources to help you learn and understand all the different genres involved in the creative industry of illustration, graphic design, concept art, and digital matte painting. Continue to learn, explore, and decide on what you want to focus on first to build up your skills, and then branch off from there. Let's move on to the final section.

Summary

It was a long stretch starting with *Chapter 1, Taking Your Design and Editing to the Next Level with Photopea*, leading up to this final chapter. You got a taste of different things you can achieve using Photopea throughout this book as a foundation.

One of the key features we discussed was Smart Objects, which allow for non-destructive editing and manipulation of images.

Aside from smart objects, we also touched upon lesser-known features such as the Vanishing Point filter, which can be a game-changer for creating realistic perspectives in your designs.

Layer Comps were another highlight, showcasing how you can easily manage different versions of your artwork in a single document. We did a step-by-step tutorial showing how layer comps offer a versatile way to show different variations and arrangements of an illustration and are invaluable for creating logos, page layouts, and other graphic design projects. In addition, we covered the basics of adding animations to images and drawings, opening up new possibilities for dynamic visual storytelling.

Lastly, we emphasized the importance of tailoring your portfolio to showcase the type of work you are passionate about. By curating your portfolio to align with your interests and career goals, you can make a lasting impression on potential clients or employers.

By incorporating these tips, tricks, and best practices into your design workflow, you can elevate your skills and create compelling visuals that resonate with your audience.

In closing this last chapter, I would like to thank you for choosing this book and being open to learning from my creative processes and experiences. Preparing this book to teach others has revealed not only my strengths but also some areas I could, and did, improve in.

Although we covered quite a lot of topics and examples in this book, it was still impossible to cover everything about Photopea and all there is to know about image editing in one book.

You will still have a solid foundation in both photo editing knowledge and using Photopea and similar programs like it to further your knowledge, skills, and understanding over time.

Congratulations on reaching the end of this book! Image editing can cross over into many different creative backgrounds, and since creative technology is always improving, there will always be something new to learn.

I look forward to hearing from you and seeing what you create.

Index

R

Raster images 9
versus Vector images 10
raster layer 69
raster mask 81
creating 81-83
red channel
creating 347
Refine Edge tool 121
using 121, 122
requests for proposal (RFPs) 272
RGB 11
versus CMYK color modes 11
ruler tool 29-31

S

sans serif fonts
anatomy 272
usage 272
versus serif fonts 271
selection 95
making 95, 96
making, with Ellipse Select tool 99-101
making, with Shape tool 99
selection tools, using 96, 97
shapes, creating 99
serif fonts
anatomy 272
usage 272
sidebar 62
Smart Objects 429
benefits 430
creating 430-433
working with 433-440
Snapping feature
using 32

sRGB 154
standard color spaces 155
CIE 1931 155
CIELAB 155
CIEXYZ 155
style guides 442
swatches 144, 145
backing up, with layers and shapes 151-153
colors, applying to shapes and images 145
colors, deleting 147
importing, into panel 150
managing 146

T

templates 300
creating 16
text
adjusting, with Character panel 288-291
filling, with image 292-297
Text styles 280
character styles 281
custom fonts 286
custom fonts, downloading 286
paragraph styles 283-285
working with 273
Text tool 273, 276, 277
paragraph text 274
point text 274
text on curve 275
Type Layer, editing 278, 279
working with 273
toolbar 46
Blur Tool 55
Brush Tool 53
Burn Tool 56
Clone Tool 54
Content-Aware Move Tool 52

V

W

Y

www.packtpub.com

Subscribe to our online digital library for full access to over 7,000 books and videos, as well as industry leading tools to help you plan your personal development and advance your career. For more information, please visit our website.

Why subscribe?

- Spend less time learning and more time coding with practical eBooks and Videos from over 4,000 industry professionals

- Improve your learning with Skill Plans built especially for you

- Get a free eBook or video every month

- Fully searchable for easy access to vital information

- Copy and paste, print, and bookmark content

Did you know that Packt offers eBook versions of every book published, with PDF and ePub files available? You can upgrade to the eBook version at packtpub.com and as a print book customer, you are entitled to a discount on the eBook copy. Get in touch with us at customercare@packtpub.com for more details.

At www.packtpub.com, you can also read a collection of free technical articles, sign up for a range of free newsletters, and receive exclusive discounts and offers on Packt books and eBooks.

Other Books You May Enjoy

If you enjoyed this book, you may be interested in these other books by Packt:

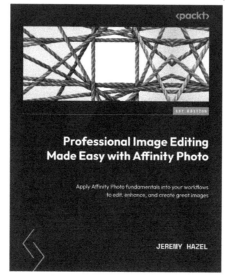

Professional Image Editing Made Easy with Affinity Photo

Jeremy Hazel

ISBN: 978-1-80056-078-9

- Develop a repeatable methodology for use in your photo editing style
- Understand the most frequently utilized techniques by professional editors
- Compete a body of work for use and reference in future projects
- Develop your own libraries of assets, macros and tone mapping presets for your style
- Recreate popular editing styles utilized in print and film
- Recondition older family photos
- Understand and deconstruct other editing styles to expand your knowledge

Mastering Adobe Photoshop 2024

Gary Bradley

ISBN: 978-1-83882-201-9

- Discover new ways of working with familiar tools, enhancing your existing knowledge of Photoshop
- Master time-saving retouching techniques, ensuring flexibility for repeated edits without compromising on quality
- Create precise image cut-outs and seamless montages with advanced masking tools
- Make Photoshop your go-to application for social media content
- Automate repetitive tasks with actions and scripts that batch-process hundreds of images in seconds
- Integrate vector assets, type styles, and brand colors from other CC applications

Packt is searching for authors like you

If you're interested in becoming an author for Packt, please visit authors.packtpub.com and apply today. We have worked with thousands of developers and tech professionals, just like you, to help them share their insight with the global tech community. You can make a general application, apply for a specific hot topic that we are recruiting an author for, or submit your own idea.

Share Your Thoughts

Now you've finished *Unlock Your Creativity with Photopea*, we'd love to hear your thoughts! Scan the QR code below to go straight to the Amazon review page for this book and share your feedback or leave a review on the site that you purchased it from.

https://packt.link/r/1-801-81664-6

Your review is important to us and the tech community and will help us make sure we're delivering excellent quality content.

Download a free PDF copy of this book

Thanks for purchasing this book!

Do you like to read on the go but are unable to carry your print books everywhere?

Is your eBook purchase not compatible with the device of your choice?

Don't worry, now with every Packt book you get a DRM-free PDF version of that book at no cost.

Read anywhere, any place, on any device. Search, copy, and paste code from your favorite technical books directly into your application.

The perks don't stop there, you can get exclusive access to discounts, newsletters, and great free content in your inbox daily

Follow these simple steps to get the benefits:

1. Scan the QR code or visit the link below

https://packt.link/free-ebook/9781801816649

2. Submit your proof of purchase
3. That's it! We'll send your free PDF and other benefits to your email directly

www.ingramcontent.com/pod-product-compliance
Lightning Source LLC
Chambersburg PA
CBHW060110090326
40690CB00064B/4500